COVID-19 & Travel

Impacts, Responses and Outcomes

Simon Hudson

(G) Goodfellow Publishers Ltd

(G) Published by Goodfellow Publishers Limited,
26 Home Close, Wolvercote, Oxford OX2 8PS
http://www.goodfellowpublishers.com

British Library Cataloguing in Publication Data: a catalogue record for this title is available from the British Library.

Library of Congress Catalog Card Number: on file.

ISBN: 978-1-911635-71-0

DOI: 10.23912/9781911635703-4387

Copyright © Simon Hudson, 2020

All rights reserved. The text of this publication, or any part thereof, may not be reproduced or transmitted in any form or by any means, electronic or mechanical, including photocopying, recording, storage in an information retrieval system, or otherwise, without prior permission of the publisher or under licence from the Copyright Licensing Agency Limited. Further details of such licences (for reprographic reproduction) may be obtained from the Copyright Licensing Agency Limited, of Saffron House, 6–10 Kirby Street, London EC1N 8TS.

All trademarks used herein are the property of their repective owners, The use of trademarks or brand names in this text does not imply any affiliation with or endorsement of this book by such owners.

Design and typesetting by P.K. McBride, www.macbride.org.uk

Cover design by Cylinder

Cover photo by Bruno Cervera on Unsplash

Contents

Acknowledgements

I am grateful to the many individuals and organizations that helped to make this book *COVID-19 & Travel* a reality. In particular, I would like to thank Sally North, Mac McBride and Tim Goodfellow from Goodfellow Publishers for their professional support throughout the writing process. I also appreciate the editorial help my wife Louise gave me, and the hard work of University of South Carolina students Caitlin Ansel and Niki Berlinsky. Finally, the book has benefited tremendously from the people in the industry who took the time to talk to me during a difficult period. These people are too numerous to name individually, but their enthusiasm and passion for travel was obvious in communications with them, and I am indebted to their contribution.

Preface

The travel industry worldwide has been dealt a vicious blow by COVID-19. The industry will recover, but travel will never be the same again, and the year 2020 will be a defining moment in the history of the tourism sector. As COVID-19 has painfully demonstrated, travel can play a critical role in the spread of new infectious diseases. Likewise, the increased globalization of tourism means that the industry is uniquely vulnerable to the disruption these disasters can cause. Despite occasional shocks, tourism has experienced continued expansion and diversification since the 1950s to become one of the world's largest and fastest-growing economic sectors. Because of this growth, travel and tourism may be the one industry to see the greatest impact from the COVID-19 crisis, and so it is critical that this event be documented.

COVID-19 & Travel: Impacts, Responses and Outcomes is divided into six chapters, with each chapter beginning and ending with a case study that reflects the material covered in the text. The first chapter tracks the period between the first signs of the virus at the end of 2019 to the beginning of April 2020, showing how as COVID-19 spread, so too did its impact on the travel sector around the world. The opening case study highlights how the cruise industry suddenly went from being the golden child of the tourism sector to a sober symbol of the deadly disease, and second case study analyzes the meltdown of ski resorts around the world as the virus tightened its grip on the tourism industry in March 2020.

By mid-April 2020, a third of the global population was under full or partial lockdown, and Chapter 2 documents this period, a time when the travel sector worldwide continued to experience a loss of business. The chapter analyzes the of challenges of leadership during the pandemic, with a case study about Richard Branson epitomizing the pitfalls. How the travel industry adapted to lockdown is discussed in this chapter, with the final case study focusing on hotels that pivoted during lockdown to lend a helping hand.

Chapter 3 discusses communication during the crisis, and how some organizations understood the importance of keeping lines of communication open – and others didn't. The case studies in this chapter profile *Micato Safaris* and *Auckland Tourism, Events and Economic Development* – two organizations that clearly understood the importance of communicating during a crisis. The chapter analyzes the various communications strategies used by the travel sector during the pandemic, including how some organizations used cause-related marketing to portray an image of corporate responsibility. The final section of the chapter emphasizes the importance of internal communication during a crisis.

The focus of Chapter 4 is the consumer. At the beginning of the COVID-19 outbreak, thousands of travelers had travel plans disrupted, and many were stranded abroad. This chapter looks at the consequences of such disruptions, but also touches on a certain segment of travelers who were oblivious to the crisis – either due to a lack of knowledge or to a lack of common sense. Consumer behavior during the lockdown period is examined, followed by a synopsis of the research undertaken during this period concerning future travel behavior, as the industry sought to understand who would travel first and when, once lockdown regulations were eased. The opening case study follows the trials and tribulations of two cruise passengers during the outbreak, and the final case describes how VisitBritain maintained a dialogue with consumers during lockdown.

Chapter 5 looks in more detail at the economic, social and environmental impacts of COVID-19. Most of the studies to date about the consequences of the pandemic have emphasized the economic impacts, and a synopsis of those studies are provided. However, there have been significant social and environmental impacts resulting from the crisis that have affected the travel sector, so these are also discussed in this chapter. The opening case study profiles Aruba, a tourism destination extremely dependent on tourism, and the end-of-chapter case study explores the various impacts of the crisis on Italy, one of the worst-affected nations.

Until there is vaccine, COVID-19 will influence nearly every sector of travel industry, so Chapter 6 takes a peek into the future for the different industry sectors, a future that will be heavily influenced by technology and a heightened emphasis on health and safety. A section in this chapter focuses on a theme that has been prevalent in this book – the need for adaptability or 'COVID-aptability'. Consumer demands and behavior will be permanently altered by the pandemic, and all stakeholders in the travel industry will need to adapt. One part of adaptability is redesigning servicescapes – a necessity for many after the lockdown – and this is the subject of the penultimate section of the chapter. The conclusion looks at the important lessons learned from this crisis for those in the travel sector. Vietnam is the subject of the opening case study, and the final case study looks at how one entrepreneur in Canada is doing his best to survive the 'new normal'.

I am a big fan of historian, author, and world-traveler Yuval Noah Harari, who wrote *Sapiens* and *Homo Deus*. At the end of this book I noted that Harari believes historians in a thousand years will look back at the COVID-19 crisis as a mere bump in the road for the human race. However, those of us involved in the travel industry will see this crisis more as a road block. So this book is intended to provide a few ideas for navigating around this road block, and for being prepared and more resilient when the next one comes along.

Simon Hudson

About the author

Simon Hudson is a part-time professor at the University of South Carolina and a consultant for the travel industry. He has written 10 books, and over 100 research articles and book chapters. With an eclectic background in the ski industry, retail, and academia, Simon is a fount of international experience and comprehensive business information. His cosmopolitan and creative ideas have been influenced by award-winning work at the University of Calgary and the University of Brighton, as well as visiting academic positions he has held in Austria, Switzerland, Spain, Fiji, New Zealand and Australia. Simon has also taught three times on Semester at Sea, a floating university that circumnavigates the world. You can reach Simon at shudson@hrsm.sc.edu

1 An unfolding crisis

Case study: How the cruise industry became a symbol of COVID-19

Figure 1.1: The Diamond Princess (courtesy of Princess Cruises image library)

It was February 3, 2020, and passengers aboard Carnival's Diamond Princess were enjoying the last night of their two-week cruise. Yes, they had heard about a mysterious virus that was scything through mainland China, but it looked like it wouldn't be affecting their vacation and they would be arriving in Japan the next day. They tucked into what they thought would be their last meal on the ship, unaware that an 80-year-old passenger – a man who had coughed through the first third of the cruise – had been admitted to a hospital in Hong Kong. Suddenly over dinner the ocean liner's intercom came to life, and the ship's captain told everyone about the passenger in Hong Kong who had tested positive for the new virus. Accordingly, he said, when the Diamond Princess reached Yokohama, everyone would need to stay on the vessel for an extra day while Japanese health officials screened them.

By morning, as the ship idled close to shore, nine more passengers and one crew member had tested positive for the coronavirus. All passengers were to return to their cabins, where they would remain quarantined for two weeks by order of the Japanese government. Rather than release 3,700 potential vectors – who could infect Japan or their homes countries – public health officials were transforming the Diamond Princess into a floating quarantine center. On the second day of the quarantine, the captain announced that the number of cases had doubled to 20. By day nine, when it reached 218, the ship had more cases than every nation in the world except China. In late March, the Centers for Disease Control and Prevention (CDC) reported that out of the Diamond Princess's 3,711 passengers and crew, 712 had eventually tested positive. Nine people had died. These numbers were a fraction of the sizeable casualties steadily accumulating across the globe, but from the Diamond Princess emerged a few of the original germs from which a huge tragedy would grow.

Until COVID-19, the cruise sector was one of the most successful and dynamic subsectors of the global tourism industry. The number of passengers carried by the cruise industry had grown year-on-year and was about 30 million in 2019, responsible for over a million jobs and over $50 billion in wages and salaries. Many destinations relied heavily on the cruise sector for their income. Passengers spend on average $376 in port cities before boarding a cruise and spend $101 in each visiting port destination during a cruise. Around 32 million passengers were expected to set sail in 2020, and to meet ongoing demand, CLIA Cruise Lines were scheduled to debut 19 new ocean ships in 2020.

But then along came the pandemic, and the cruise industry suddenly went from being the golden child of the tourism sector to a sober symbol of the deadly disease. Cruise ships were a focal point of the pandemic from the beginning, widely blamed for a series of major outbreaks that helped spread the disease across the world. No cruise operator was hit harder than Carnival. At least seven of the company's ships at sea became virus hot spots, resulting in more than 1,500 positive infections and at least 39 fatalities. Carnival's Princess Cruises mentioned above gained most of the attention, drawing worldwide attention and leading to several countries repatriating their citizens from the ship. But two other Carnival cruise ships were at the center of the outbreak.

Shortly before the coronavirus was declared a pandemic, and with over 2,700 passengers on board, another ship from the Princess Cruises portfolio, the Ruby Princess, sailed into international waters despite a global increase of confirmed cases of COVID-19. By mid-April, there were 662 confirmed cases among Australian passengers alone and 19 deaths. The subsequent discharge of infected passengers into Australia worsened the national pandemic there to the extent that on April 5, New South Wales police announced they had launched a criminal investigation into whether the operator of the Ruby

Princess downplayed potential coronavirus cases before thousands of passengers disembarked in Sydney.

If this wasn't a big enough blow for Carnival, another of its cruise ships, the Grand Princess, was also in the headlines due to the coronavirus. Early in March, American Vice President Mike Pence announced that 21 people on board the ship off the coast of California had tested positive for the coronavirus. The panic over the fate of the ship, which was returning from Hawaii, began when a 71-year-old man died after traveling on a previous leg of the cruise, a round trip from San Francisco to Mexico. Then, more cases with links to the ship emerged. In Placer County, northeast of San Francisco, officials announced three new cases – all passengers who had previously traveled on the Grand Princess on a trip to Mexico. Nearby in Contra Costa County, officials also announced three new COVID-19 cases, including two who had been aboard the Grand Princess.

Carnival's response to the coronavirus outbreak raised questions about corporate negligence and fleet safety, and as a result, US Congress opened an inquiry into the company's handling of the outbreak. President and Chief Executive Officer Arnold Donald said his company's response was reasonable under the circumstances. "This is a generational global event – it's unprecedented," he said. Carnival canceled all its cruises in mid-March, and the company's share price fell sharply, along with the rest of the industry. Despite assurances from President Trump, the cruise sector was left out of his country's $2 trillion stimulus package, so at the end of March, Carnival began an effort to raise $6 billion by selling stock, bonds and other securities. Donald said the sale would generate enough cash for the company to survive without revenue into 2021. He added that Carnival hoped to take advantage of stimulus programs in other countries where it operates, like Germany, Britain and Australia. Before the coronavirus crisis began, the company's nine cruise brands employed 150,000 people who served nearly 11.5 million travelers a year, a significant segment of the global market.

Carnival and the cruise industry will recover. Unlike many restaurants and other smaller tourism business that are now on the brink of collapse, the cruise lines started the crisis with healthy balance sheets, and analysts say the major cruise lines have enough money to survive at least six months on lockdown. Several cruise lines turned to social media as a means of maintaining interest with their loyal consumer base. Using videos and virtual tours, Viking helped homebound travelers to tour popular destinations, while Carnival crewmembers posted pictures from the ships. Holland America streamed a Lincoln Center musical performance from its onboard entertainment, and a former Royal Caribbean International cruise director even created a virtual cruise with daily installments, all to remind travelers of the joys of cruise travel. "The cruise industry will weather this outbreak," concludes Rhonda Weaver, a travel agent based in Kent, Washington. "But it will take time and a lot of work."

Sources: Carr & Palmeri (2020); Yaffe-Bellany (2020); Jordan (2020); Clark (2020)

Introduction

The travel industry worldwide has been dealt a vicious blow. It is forecast that the number of international tourist arrivals will fall by 60-80%% in 2020 due to the novel coronavirus, putting millions of jobs at risk (Alpert & Beilfuss, 2020; UNTWO, 2020). The drop in arrivals will lead to an estimated loss of $300-450 billion in international tourism receipts (The Economic Times, 2020). The industry will recover, but travel will never be the same again, and the year 2020 will be a defining moment in the history of the tourism sector. But how did this crisis unfold and start to impact travel? This chapter will track the period between the first signs of the virus at the end of 2019 to the beginning of April 2020, showing how as the virus spread, so too did its impact on the travel and tourism around the world.

As the opening case study demonstrated, the 2020 pandemic and travel were inextricably linked. As COVID-19 has painfully demonstrated, travel can play a critical role in the spread of new infectious diseases. The ability to get to nearly any country in the world in 20 hours or less, and pack a virus along with our carry-on luggage, allows new diseases to emerge and to spread when they might have died out in the past (Walsh, 2020b). Likewise, the increased globalization of tourism means that the industry is uniquely vulnerable to the disruption these disasters can cause (Hudson & Hudson, 2017). Travel and tourism may be the one industry to see the greatest impact from the coronavirus (see Figure 1.2), and it may take many years for the sector to recover.

Despite occasional shocks, tourism has experienced continued expansion and diversification since the 1950s to become one of the world's largest and fastest-growing economic sectors. Until recently, international tourism represented 7%

Industry impact analysis

| Covid-19 Industry Impact
boardofinnovation.com | Industry characteristics | | | | | |
| | 'If characteristic is present in your or your clients' business, impact is negative (unless you successfully pivot). | | | | | |
Note: Detailed impact analysis per industry in dedicated reports.	Large gatherings are essential	Close human interaction is essential	Hygiene, or perception thereof, is critical	Dependant on travel (business and leisure)	Service or product is postponable or expendable	Impact analysis
Tourism and hospitality	Very high	Very high	High	Very high	High	Very high
Sports	Very high	Very high	Medium	Low	Medium	High
Music	High	High	Low	Medium	Medium	High
Automotive	Low	Low	Medium	Low	Very high	High
Beverages (Alcohol)	High	High	Medium	Medium	Low	Medium
Retail (non-food)	High	Medium	Medium	Medium	Medium	Medium
Pharmaceuticals	Low	Low	High	Low	Low	Low

Figure 1.2: Industry impact analysis from COVID-19 (courtesy of Board of Innovation, 2020: 26)

of total world exports and 30% of services' exports. The impact of tourism goes far beyond enrichment in purely economic terms, helping to benefit the environment and culture and the fight to reduce poverty. Over the past decade, the annual growth rate of tourists visiting developing countries has been higher than the world average growth rate in tourism. In many of these countries, tourism can serve as a foothold for the development of a market economy where small and medium-sized enterprises can expand and flourish. And in poor rural areas it often constitutes the only alternative to declining farming opportunities.

The outbreak

At the end of December 2019, public health officials from China informed the World Health Organization (WHO) that they had a problem: an unknown, new virus was causing pneumonia-like illness in the central Chinese city of Wuhan. WHO named the disease caused by the virus COVID-19, which references the type of virus and the year it emerged. Coronaviruses are common in animals, but they sometimes evolve into forms that can infect humans. Since the start of the century, two other coronaviruses have jumped to humans, causing the SARS outbreak in 2002, and the MERS outbreak in 2012. COVID-19 moved rapidly around the world, spreading particularly quickly in contained environments, like on the cruise ship the *Diamond Princess.*

On January 20, 2020, China confirmed human-to-human transmission of the virus after medical staff in Guangdong Province were infected. The country took aggressive action at the start of the outbreak, shutting down transportation in some cities and suspending public gatherings. Officials isolated sick people and aggressively tracked their contacts, and had a dedicated network of hospitals to test for the virus. On January 21, WHO called the coronavirus situation 'an emergency in China' but stopped short of declaring a 'public health emergency of international concern'. This designation was created after the 2003 SARS outbreak, and five have been declared since 2009: swine flu, polio, Ebola, Zika and novel coronavirus. The next day, Europe's first three coronaviruses cases were confirmed in France. On January 22, WHO declared a global health emergency. Inside China, the virus had infected nearly 8,000 and killed at least 170. It had also spawned 98 cases in 18 other countries.

■ Ripple effects

It was early February when COVID began to have a potent impact on the travel sector. As the opening case study highlighted, Carnival's *Diamond Princess* cruise ship hit the headlines on February 3 when it was quarantined off the coast of Japan following a confirmed outbreak of the coronavirus on-board. Around 3,700 people, including 2,600 passengers and 1,000 crew members were quarantined on the ship for nearly a month. More than 700 passengers and crew members ultimately tested positive for the virus. Following in the footsteps of the Diamond

Princess, at least 25 other cruise ships confirmed COVID-19 cases — including 78 cases on the Grand Princess, which was quarantined off the coast of California.

The airline sector was also affected at this time, and as airlines restricted flights in and out of China and large areas were placed into quarantine, Chinese travelers started to cancel trips abroad. Chinese travelers racked up an unprecedented 150 million trips abroad and spent more than $277 billion on international travel in 2018, so these cancelations did not go unnoticed. The Asian countries closest to China felt the brunt of the crisis. Vietnam, Thailand, Cambodia, Malaysia and Singapore were each expected to lose at least $3 billion in tourism-related revenues, according to an analysis by Animesh Kumar, a travel and tourism director at GlobalData, a London-based research and consulting firm (Scott, 2020).

But the ripple effects of the outbreak in China were spreading beyond Asia. In Northern Ireland, for example, 3,000 hotel rooms were canceled by Chinese tourists between January and March. Colin Neill, of Hospitality Ulster, said the cancelations were "causing real concern". He said Chinese tourists were a "really significant element of our tourism" and that the cancelations were a "huge challenge to the industry" (Morgan, 2020). The cancellation of trips in the country had an effect on coach companies in particular. Caroline McComb, of McComb's Coaches said "The entire tourism industry in Northern Ireland is obviously going to be affected by this and because Asia is such a growth market for us we really do need to work hard to develop other markets" (Morgan, 2020). Tourism Ireland estimates 100,000 Chinese tourists visited Ireland in 2018.

China itself was naturally suffering from an increasing number of cancelations from inbound travelers at this time. "The numbers of trip cancelations – not just to China but to the entire continent of Asia – is growing every day," said Jack Ezon, founder and managing partner of luxury travel agency Embark Beyond. "People are put off. Sadly, a lot of them are just saying, 'I don't know if I want to go anywhere right now'. Or, in many cases, 'I'll just go next year'." By mid-February, almost 75% of his travelers had canceled their February and March departures to Southeast Asia, which the US Centers for Disease Control and Prevention still considered at that time to have a lower, level one risk for the coronavirus. "They're worried about being anywhere close to the outbreak," he said. "Or of getting stuck with canceled flights if other hubs become infected" (Ekstein, 2020).

In Europe in the last week of February, with the news that 11 towns in Italy were on lockdown, and countries like Austria and Croatia had announced their first cases, it was apparent that the impact of COVID-19 was likely to be felt on a more global scale than was perhaps previously envisaged. "If there was previously a temptation to view the coronavirus as a China, or Asia, issue then developments this week must force a shift in mindset," said Nick Wyatt, head of R&A, Travel and Tourism at GlobalData (Choat & Wilson, 2020). The Venice Carnival was cut short on February 23; Milan Fashion Week limited public access to shows; and 40 football matches, including four Serie A games, were postponed, along with dozens of church services, walking tours and opera performances.

Figure 1.3: Italy was the first European country to lockdown (photo by Victor He on Unsplash)

Elsewhere in Europe, guests were confined to rooms at a hotel in Tenerife after an Italian doctor and one guest who stayed at the hotel tested positive for the virus; and a hotel in Innsbruck, Austria, was in lockdown after its Italian receptionist tested positive. However, the majority of airlines servicing Europe were still operating routes as normal, and most tour operators said that normal booking conditions still applied. "No one has any idea how soon the coronavirus will be brought under control, so it's a waiting game at present. Most Association of Independent Tour Operators (AITO) with destinations directly affected are running on a rolling three-week plan: they are not even looking at canceling or amending any booked departures more than three weeks ahead," said Derek Moore, chairman of AITO (Choat & Wilson, 2020).

By late February, as cases of the coronavirus continued to multiply in China, and concerns about the disease led travelers to cancel upcoming trips to Asia, tourism officials elsewhere saw an opportunity. In Alaska, for example, officials with Ted Stevens Anchorage International Airport and Visit Anchorage, the tourism marketing organization for Alaska's largest city, began lobbying airlines, travel agents and tour operators to increase airline service, reroute cruises and generally get the word out about the sights and attractions of the northernmost state. "Tour operators that were selling tour packages into Asia are seeing significant cancelations because of concerns about coronavirus, but people with those packages still want to travel somewhere," said Jim Szczesniak, manager at the Anchorage airport. "What we're working on is attracting the demand from those people who want to travel." The marketing initiative was geared toward travelers from Australia, Northern Europe and the continental United States. "We just want people who were going to China to think of Alaska as a temporary replacement for a trip they can rebook to China in the future," said Julie Saupe, president and chief executive of Visit Anchorage (Mzezewa, 2020).

Alaska wasn't the only tourist destination looking to lure tourists. A new advertising campaign called *There's Still Nothing Like Australia* attempted to convince Americans and Britons that Australia was a safe alternative to Asia. Tourism Minister Simon Birmingham said the government was looking at targeting tourists still willing to travel and emphasizing that Australia was a safe location despite the threat of coronavirus elsewhere. He said in February that Australia was in a "strong position in terms of being seen as a very safe destination for those who might be concerned about coronavirus" with no new recent infections in the country (Duke & Bagshaw, 2020).

Elsewhere, in Japan, an *Empty Kyoto* campaign promised travelers that if they visited Kyoto they would have some of the most popular locations to themselves. Merchants from five shopping streets in Kyoto's Arashiyama neighborhood – a popular tourist district on the western outskirts of the city that's filled with temples and shrines – devised the advertising campaign dubbed "suitemasu Arashiyama," which translates to "empty Arashiyama" or "there are few people around in Arashiyama." The posters created for the campaign showcased how any would-be travelers could have the district's most-visited spots all to themselves. The coronavirus outbreak in China particularly affected Japan, which had approximately 9.6 million Chinese visitors in 2019 – a third of foreign tourist expenditure in the country (Jozuka, 2020).

The Philippines, on the other hand, promoted safety as opposed to no crowds. The Department of Tourism (DOT) announced in early March that it would be spending some six billion pesos to boost the tourism industry, which was heavily hit by travel bans from China, Hong Kong and Macau. The resiliency program focused on promoting domestic destinations and on assuring the public that it was safe to travel within the Philippines. In a Facebook video, President Rodrigo Duterte encouraged people to travel domestically because "everything is safe in our country" (Mzezewa, 2020). The DOT planned for the President to visit top tourism destinations like Boracay, Cebu, and Bohol in order to promote domestic tourism.

Pandemic

The crisis struck during the key pre-summer booking period for many travel companies, prompting a rush to find ways to reassure prospective customers. Early in March, for example, British Airways unveiled a new "book with confidence" policy: any bookings made before March 16 could be postponed, or the route changed, without paying a fee. The Spanish-based Melia hotel group, with more than 380 properties worldwide, ran a flash sale that combined deep discounts of up to 45% with the option to cancel at no charge for the rest of the year. Cruise lines took similar steps. Windstar, for example, was allowing cancelation up to 15 days before setting sail (but for a credit note rather than a refund). Princess Cruises offered bookings for 2020 and 2021 on payment of a deposit of just £1 (Robbins, 2020). Canadian airline WestJet's discount brand Swoop was still selling one-way tickets to Mexican destinations for $101 and to Las Vegas for as low as $69, depending on departure dates and locations. "Prices so low you gotta go!" a Swoop ad said, above an image of a toilet paper roll, a cheeky reference to the shortage of bathroom tissue caused by panic buying across the country as worried people stocked their pantries in preparation for possible quarantines to avoid catching or spreading COVID-19.

But despite reassurances, travel disruptions began to be commonplace. On March 14, for example, Jet2 planes heading to Spain turned back in mid-air as the airline canceled all flights to the country. The airline, which flew from nine UK airports to destinations including Alicante, Málaga and Lanzarote, said it was canceling all flights to mainland Spain, the Balearic Islands and the Canary Islands immediately. Jet2 said in a statement: "In response to local measures introduced throughout Spain to prevent the spread of COVID-19, including the closure of bars, restaurants, shops and activities including any water sports, we have taken the decision to cancel all flights ... with immediate effect" (Skopeliti, 2020).

COVID-19: Unprecedented Decline In Air Traffic

Number of flights tracked daily worldwide
(01 Feb-30 Mar, 2020)

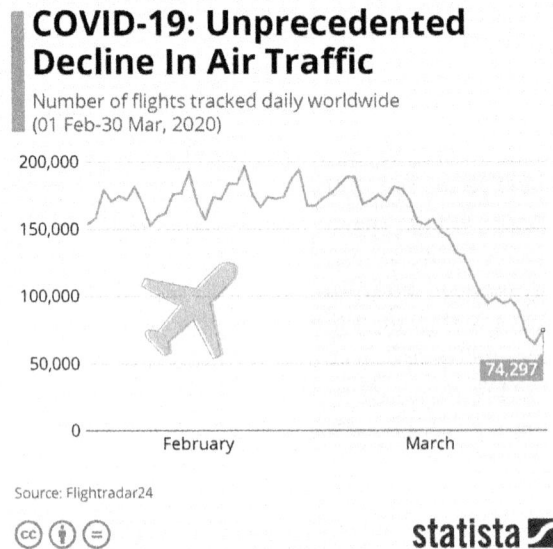

Source: Flightradar24

statista

Figure 1.4: Decline in air traffic from February 1 to March 30, 2020 (courtesy of Statista)

Flight tracking data from website Flightradar 24 shows the scale of disruption COVID-19 caused the global airline industry (see Figure 1.4). The data shows that 196,756 flights were tracked on February 21, a number that tumbled to approximately 155,000 around the middle of March. By March 29, it had fallen even further to just 64,522 (McCarthy, 2020). Cruise ships were also left stranded. On March 13, in light of mounting fears over onboard COVID-19 outbreaks, Cruise Lines International Association (CLIA) made the decision to suspend operations from US ports of call for 30 days. Two weeks later, thousands of passengers and crew members remained aboard at least 15 ships across the world.

Interestingly, one of the countries that kept the virus under control when other parts of the world had not was Taiwan, and they made a very early decision (the beginning of February) to stop cruise ships docking at the island's ports – in addition to banning travel from many parts of China (Griffiths, 2020a). The severe acute respiratory syndrome (SARS) outbreak of 2003 sent shockwaves through much of Asia and cast a long shadow over how people responded to future outbreaks. This helped many parts of the region, including Taiwan, react faster to the coronavirus outbreak and take the danger more seriously than others, both at a governmental and societal level. Border controls and the wearing of face masks became routine as early as January in many areas.

The resort sector was not immune to the sudden challenges faced by others in the tourism industry. Once again, Asia felt the first shockwaves with many resorts closing in early March. China, where the outbreak originated and where most cases were originally reported, makes up a significant portion of sales for resort companies. Casino and resort company Las Vegas Sands earns 90% of its income in China, and Wynn Resorts earns 73% of its income there (Miranda & Atkinson,

2020). As the month progressed, resorts in other parts of the world started to close. In the Caribbean, for example, Sandals and Beaches resorts shut down operations in the last week of March. The all-inclusive resort chain said none of its 15 Sandals resorts nor three Beaches properties would accept any guest arrivals past March 23. Many ski resorts in Europe and North America, just before the Easter crowds were due to arrive, were also forced to shut down operations in March. The case study at the end of the chapter looks at how this unfolded in more detail.

In the attractions sector, Shanghai Disney was among the first major global attractions to close its doors – that was in late January – followed in short order by all of Asia's Disney-themed parks. In the US, Disney announced on March 12 that it was closing Walt Disney World in Orlando, Florida, as well as its Disneyland resort in Los Angeles. At the same time, the company suspended all new departures with the Disney Cruise Line. Universal Studios theme parks in Orlando and Los Angeles also announced plans to temporarily close. Waterparks globally were forced to close due to restrictive measures put in place by respective governments. While this mostly affected indoor waterpark resort operations, it also came at a time when outdoor waterparks were readying to open for the 2020 summer season.

Figure 1.5: Hong Kong Disneyland closed at the end of January (photo by Kon Karampelas on Unsplash)

The same day Disney announced its closures in the US, Broadway shows were suspended by The Broadway League, a national trade association. The move came in "support of the health and well-being of the theatre-going public, as well as those who work in the theatre industry," the association said. "Our top priority has been and will continue to be the health and well-being of Broadway theatregoers and the thousands of people who work in the theatre industry every day, including actors, musicians, stagehands, ushers, and many other dedicated professionals," said Charlotte St. Martin, president of the Broadway League (Romine, Levenson & Morgado, 2020).

In the hotel sector, not surprisingly, China was the first country to witness cancelations due to COVID-19. Hotel occupancy declined about 75% in Mainland China from January 14-26, so the Chinese New Year holiday period at the end of January was significantly impacted by the outbreak. By mid-February, the travel setback caused by the coronavirus had spread beyond China, with other parts of the Asia Pacific region experiencing a slowdown in outbound travel bookings. By the end of that month, travel in Asia was almost at a standstill. Macau, for example, fell from 96% occupancy to just 3% in a matter of weeks. The shutdown of casinos in the market was a key reason behind the significant drop.

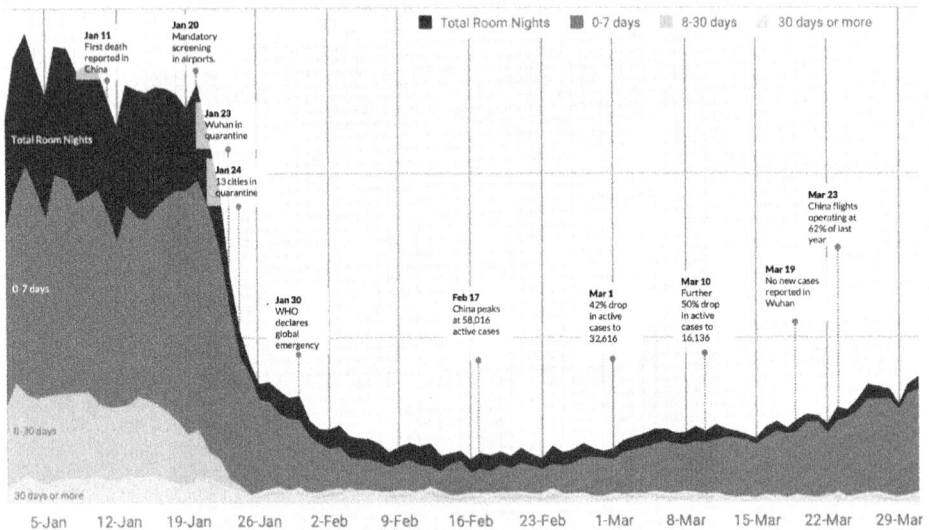

Figure 1.6: COVID-19 crisis recovery data from China (courtesy of Shiji Distribution Solutions, 2020)

Other hotels around the world that relied heavily on revenue generated from conferences and events were also beginning to suffer. Mobile World Congress 2020 in Barcelona was canceled, with the organizers saying "the global concern regarding the coronavirus outbreak, travel concern and other circumstances, make it impossible to hold the event". The conference was scheduled to take place February 24-27 and, the year before, generated about $500 million in revenue and created over 14,000 part-time jobs (Eisen, 2020). At that time, the coronavirus impact on global travel was also sending tremors through New York City's hotel industry, which was sagging from declining occupancy and revenue even before the pandemic struck (Cuozzo, 2020). Foreign visitors to the city, both for business and leisure, traditionally spend much more on average in New York than domestic tourists, with the Chinese spending the most on a per-person, per-trip basis. Visitor numbers for hotels in New York hit a record 70 million in 2019, compared with 48.8 million in 2010. But for New York, late February was just the beginning of the nightmare.

Early in March, key hotel markets in Italy – a focal point of the COVID-19 outbreak – reported significant occupancy declines. Occupancy in Milan fell to 8.5% on March 1 amid the closures of schools, gyms, museums and other major cultural attractions, including the Duomo. In Venice on the same day, just 6% of rooms in the market were occupied. The cancelation of the last days of Carnival celebrations (February 8-25) was a factor behind the significant drop. In Florence, occupancy fell to 14% and, despite being further away from the epicenter of the outbreak, Rome also experienced declines in occupancy, with absolute occupancy at 21% on March 1.

On March 11, the World Health Organization declared that COVID-19 was a pandemic, and countries began to close borders. The US suspended arrivals from 26 European countries, sending shockwaves through the travel sector in America and Europe. According to the US Travel Association, 850,000 international visitors from Europe (excluding the UK) entered the US in March of 2019, accounting for about 29% of total overseas arrivals. These visitors spent approximately $3.4 billion in the US. Meanwhile, hotels across the world began to shed employees. In Sweden, Scandic Hotels Group gave notice of termination of employment to approximately 2,000 team members, which corresponded to about half of the company's permanent employees in the country. Wynn Resorts closed Wynn Las Vegas and Encore, committing to paying all full-time employees during the closure, and Marriott announced that it would furlough about two-thirds of its 4,000 corporate employees at its headquarters in Bethesda, Maryland, as well as two-thirds of its corporate staff abroad. During the initial period, furloughed US corporate employees would receive 20% of their salary, which could be put toward health care and other costs, while corporate staff who stayed on were subject to 20% pay cuts and reduced workweeks.

By the end of March 2020, governments in many countries – for example, top tourism destinations like Spain – had ordered the closure of their hotels, and business and leisure travel was almost at a standstill. In the US, hotel occupancy in the last week of March was down from 67% in the same week in 2019 to 22.6%. The average daily rate (ADR) was down 39.4% to $79.92, with revenue per available room (RevPAR) down 80.3% to $18.05. "Year-over-year declines of this magnitude will unfortunately be the 'new normal' until the number of new COVID-19 cases slows significantly," said Jan Freitag, STR's senior VP of lodging insights. "Occupancy continues to fall to unprecedented lows, with more than 75% of rooms empty around the nation last week. 2020 will be the worst year on record for occupancy. We do, however, expect the industry to begin to recover once the economy reignites and travel resumes" (STR, 2020).

On a positive note, China's hotel sector was showing signs of recovery at this time, with daily hotel occupancy showed reaching an absolute level of 31.8% on March 28, up from a low of 7.4% during the first week of February. Shiji Distribution Solutions analyzed booking volumes and changes in lead time between booking date and arrival date in China between January 1 and March 31. They found that bookings had started to increase at the end of March by 30% a week, but booking behavior had changed – reservations were much more 'last minute'. Before the epidemic, 70% of reservations were for 0-7 days ahead and 25% were 8-30 days ahead. After the epidemic, 90% of reservations were for 0-7 days ahead and only 5% for 8-30 days ahead (Shiji Distribution Solutions, 2020).

Figure 1.7: The Venetian in Macau sits empty after closing early February (photo by Macau Photo Agency on Unsplash)

1

The hospitality industry was one of the sectors most severely impacted by the COVID-19 pandemic. Across the UK, for example, all pubs, bars and restaurants had to close their doors in order to abide by the unprecedented measures ordered by the government. Closures of restaurants caused a ripple effect among related industries such as food production, liquor, wine, and beer production, food and beverage shipping, fishing, and farming. Businesses started to look to maintain income through alternative streams, for example by changing their offering to takeout, curbside pickup, or delivery. But many pubs, cafés, restaurants, bistros and beach bars were small family businesses which were more vulnerable to the crisis. Many had to partially or even fully shut down. The National Restaurant Association in the US surveyed more than 5,000 restaurant owners and operators towards the end of March, and found that 54% of operators had made the switch to all off-premises services and 44% had to temporarily close down. "This is uncharted territory," said Hudson Riehle, the Association's senior vice president of research. "The industry has never experienced anything like this before" (National Restaurant Association, 2020). By mid-April, four out of 10 restaurants had closed, and eight million employees had been furloughed or laid off (see Figure 1.8).

March was the month when other sectors of the tourism and hospitality industry started to feel the brunt of cancelations. Due to take place March 10-14, ITB Berlin, the World's Leading Travel Trade Show was canceled, which was a symbolic blow to the travel industry. Dr. Christian Göke, CEO of Messe Berlin GmbH, said: "With more than 10,000 exhibitors from over 180 countries ITB Berlin is extremely important for the world's tourism industry. We take our responsibility for the health and safety of our visitors, exhibitors and employees very seriously. It is with a heavy heart that we must now come to terms with the cancelation of ITB Berlin 2020" (Hospitality Net, 2020). Then the Tokyo Olympics Games were postponed until 2021 as the coronavirus outbreak deepened, an unprecedented move in the 124-year history of the modern Olympics. For Japan, it is estimated that the postponement will likely cost $6 billion in lost revenue, on top of the $12 billion Japan spent in the run-up. The bulk of that will be from actual costs – things like venue maintenance, reprinting of marketing materials and hiring new volunteers – with the remainder coming from a broader blow to the economy (Nussey & Tajitsu, 2020). Around 4.5 million tickets had been sold for the Olympics, and another 970,000 for the Paralympic Games.

Stranded

By late March with the coronavirus spreading at a rapid pace, most governments had introduced strict controls on domestic and foreign travel, which left many tourists trapped abroad. Thousands were stranded in Nepal, Sri Lanka and other locations across Asia, and foreign governments stepped up their efforts to repatriate them. Some reacted faster than others. On March 17, as stringent lockdowns

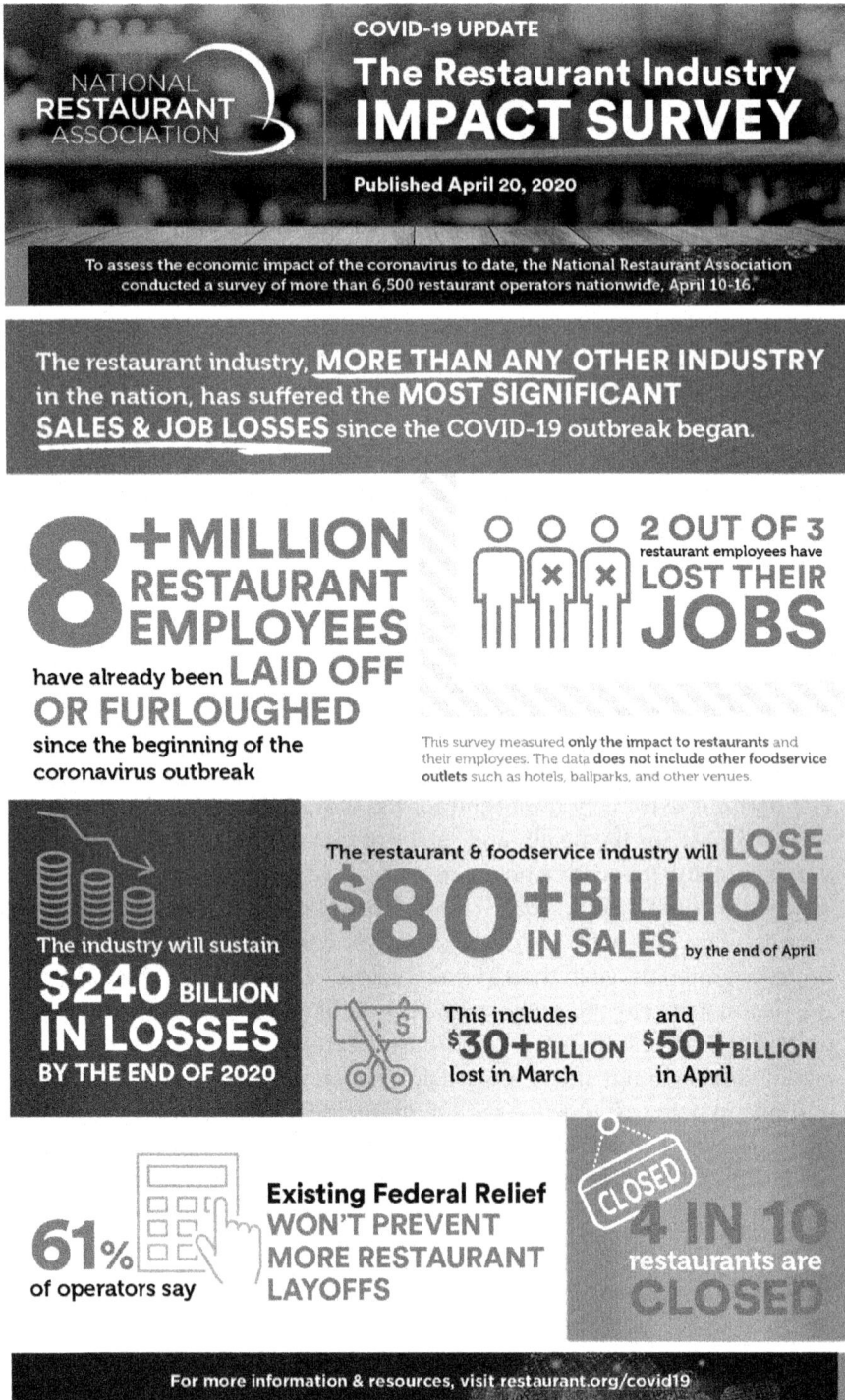

COVID-19 UPDATE

The Restaurant Industry
IMPACT SURVEY

Published April 20, 2020

To assess the economic impact of the coronavirus to date, the National Restaurant Association conducted a survey of more than 6,500 restaurant operators nationwide, April 10-16.

The restaurant industry, **MORE THAN ANY OTHER INDUSTRY** in the nation, has suffered the **MOST SIGNIFICANT SALES & JOB LOSSES** since the COVID-19 outbreak began.

8+MILLION RESTAURANT EMPLOYEES have already been **LAID OFF OR FURLOUGHED** since the beginning of the coronavirus outbreak

2 OUT OF 3 restaurant employees have **LOST THEIR JOBS**

This survey measured only the impact to restaurants and their employees. The data does not include other foodservice outlets such as hotels, ballparks, and other venues.

The industry will sustain **$240 BILLION IN LOSSES BY THE END OF 2020**

The restaurant & foodservice industry will **LOSE $80+BILLION IN SALES** by the end of April

This includes **$30+BILLION** lost in March and **$50+BILLION** in April

61% of operators say **Existing Federal Relief WON'T PREVENT MORE RESTAURANT LAYOFFS**

CLOSED

4 IN 10 restaurants are CLOSED

For more information & resources, visit restaurant.org/covid19

Figure 1.8: How the pandemic initially impacted the restaurant sector in the US (courtesy of National Restaurant Association, 2020)

started rippling around the world, the German Embassy warned its citizens of flight disruptions via its Facebook page and urged them to contact their airlines. Over the next week and a half, Germany and France organized flights to take out hundreds of people, including citizens of Finland, Austria, Denmark and Belgium. The UK government at the time pledged over $90 million to charter special flights to bring home UK nationals from countries where commercial flights were unavailable. British Airways, Virgin and Easyjet were among airlines working with the government to fly Brits back to the UK.

American tourists, though, say the US State Department was slow to inform its citizens of the challenges in returning home. Americans on three continents said in interviews that government repatriation efforts had seemed slower and less certain than those of other countries, pointing to embassy social media posts and emails through the STEP program, which provides travel updates to citizens abroad (Schultz & Sharma, 2020). Even into April, hundreds of Americans were still stranded in Peru with activists warning that a humanitarian crisis was unfolding and tourists were reportedly being evicted from hotels.

It wasn't just Peru where travelers were still stranded into April. Tourists from Australia, Britain and Canada were still stuck in India after their vacations were abruptly ended by a three-week lockdown. International flights were halted until April 14 because of the coronavirus. The Australian High Commission said it was "actively exploring options for a commercial charter flight for Australians to return. There is no guarantee and it will be difficult to achieve" (Ganguly,

Figure 1.9: Travelers were left stranded all over the world (photo by Anna Schvets on Pexels)

2020). Likewise, it wasn't until early April that thousands of Germans stranded in New Zealand due to the pandemic lockdown, were allowed to leave, although people with confirmed or suspected cases of COVID-19 were not eligible to leave the country and had to remain in isolation. New Zealand's media had featured many stories of stranded German tourists and the kindness offered to them by the citizens of New Zealand. One German couple with a baby, who were stranded in a camp site having failed to secure flights back to Frankfurt, were offered a free house to stay in by a nurse who was filling in for sick hospital colleagues in a different city (Walsh, 2020a).

Case study: COVID-19 and the meltdown of ski resorts

Mid-March and it was like a domino effect. One after the other – from St. Anton in Austria, to Verbier in Switzerland, to Aspen in Colorado – ski resorts around the world in swift succession announced closures amid growing coronavirus concerns. In the US, on the morning of Saturday March 14 – one the busiest days of the season for many ski areas – Governor Jared Polis from Colorado was looking at data on coronavirus infection rates in Colorado's ski towns, noting they were 20 to 30 times higher than average. Polis had concerns that social distancing strategies deployed by the crowded resorts were not

Figure 1.10: Sun Peaks, Canada just after closure (courtesy of author)

adequate, and wanted to shut down the ski areas. "There were many, many mountain communities that said, 'This is our livelihood. This is our business. We need the tourists.' But I'm making the decisions based on the science and the data," he said.

Melanie Mills, head of the Colorado Ski Country trade group, spoke with the governor several times that day while collecting thoughts from her 24-member ski areas. "We pushed hard for a more orderly closure than 8pm on the busiest arrival day of the year," she said. "It was a pretty intense time." However, just after 4pm that day, Vail Resorts, the largest resort operator in the continent, set off the chain reaction that would see all of Colorado's ski areas closed by nightfall and resorts across the world closed within days. CEO Rob Katz suspended operations at 34 resorts in 15 states and Canada. Minutes later, Polis issued a statement praising Vail Resorts "for taking this difficult, responsible step" and urged other resort operators to follow suit. "It was a traumatic decision for the ski industry and it was one that was made for us and not on our timetable," Mills said. "That was tough on operators, our communities where we operate and on our guests."

Your author happened to be in Sun Peaks, Canada at this time. With 75 ski days under my belt, and one of the best snow seasons on record, I thought I was heading for 100 ski days, planning to leave the day after closing date on April 13. However, the season came to an abrupt end a month early. Unlike resorts owned by Vail Resorts, at least we had a few days warning. On March 16, Sun Peaks Resort LLP (SPR) announced they would be suspending operations, effective end of day March 18, due to the spread of COVID-19. The statement they released is a good indication of how dynamic the situation was at the time:

"We will be implementing a gradual wind down of operations between now and then to help facilitate things in an orderly manner. We do not take this action lightly as skiing in the mountains was one of the safer and calming escapes people had left during this hectic time. However, the situation continues to evolve on an almost hourly basis and an appropriate response was now required. Until today, we felt our role was to continue providing an outlet for guests to recreate outdoors. As of now, we feel our role has shifted to gradually close down in an organized manner and minimize impacts for the greater good where possible…While there are no confirmed cases of Coronavirus in the Sun Peaks area at the current time, the health and safety of our guests, staff, and community is of utmost importance to our organization and destination."

European ski resorts also had little time to prepare for sudden closure. Resorts in Italy had already closed on March 9, but after the World Health Organization declared a pandemic on the 11th, Austria, France, Andorra, Italy, Norway, Spain and Switzerland all ceased operations as governments placed a limit on crowds and closed public places to stop the spread of the pandemic. France was only given four hours to shut its resorts down as the government closed anything "non-essential" to French life. French mountains officially closed at midnight on the 14th, but that very afternoon, resorts and tourist boards had reassured visitors they would stay open until the end of the season. The news, therefore, came as a surprise for holidaymakers, tour operators, airlines, and local businesses. "It's a big shock for the entire ski network," said Jean-Marc Silva, director of France Montagnes, an umbrella organization for the mountain tourism industry. "We didn't at all expect something like this to happen. The majority of the ski season is over but some stations had been planning to stay open for another six weeks. This will hit them economically."

But despite closures, the coronavirus had already ripped through well-known European ski towns. In the chic resort of Verbier, longtime destination of Europe's jet set for snow sports and après-ski high life, the packed gondola lifts and bars made it the perfect incubator for COVID-19. Even after the lifts closed, vacationers and seasonal workers partied one last time before heading home. "It's a perfect opportunity to spread – it's a virus heaven," said Corinne Cohen, a doctor at a clinic close to the resort. At one medical practice in Verbier, 80% of patients tested positive for COVID-19. Another resort in Europe, Ischgl in Austria, has been linked to hundreds of cases in Austria, Germany and Scandinavian countries, and at one time was even under investigation by Austrian authorities for negligence.

Resorts are left scrambling to deal with early closures and remain financially viable. Vail Resorts' CEO Rob Katz is taking one for the team. Going salary-less himself for six months, he is keeping the mega-ski company afloat by eliminating all cash compensation for the Board of Directors for the same period. Other measures include furloughing almost all of their US year-round hourly employees for at least a month or two (without pay but with full healthcare – a significant benefit in the US during this crisis) and implementing a six-month wage reduction for all US salaried employees. Along with these measures, he planned to cut capital costs by $80-$85 million, deferring all new chair lifts, terrain expansions and other mountain improvements, while protecting the vast majority of maintenance capital spending. At the beginning of April, Katz and his wife, Elana Amsterdam – *New York Times* bestselling author and founder of Elana's Pantry – announced a donation of more than $2.5 million to provide immediate support for both Vail Resorts' employees and the mountain towns where the company operates.

Owners of independent resorts and resident businesses that rely on ski tourism do not have such deep pockets as Vail Resorts and Rob Katz. In Sun Peaks, Canada, where I was based for the winter, early closures meant that millions of dollars that would have been spent in the resort slipped away. For 34 businesses interviewed by *Sun Peaks Independent News*, the total loss for the last four weeks of the season was about Can$4.5 million. These businesses did not include Sun Peaks Resort LLP, the Sun Peaks Grand Hotel or Bear Country, the largest employers in the village. But some of the business owners, despite major losses, say the timing of the outbreak may have saved them. "The good news is, this didn't happen in November or December," said Ryan Schmalz, owner of a local pizza restaurant.

1

Not all ski resorts around the world closed during the pandemic, and some even re-opened. China, the Czech Republic, Iceland, Sweden and Norway were all operating resorts in March and April, but with very strict opening and operating restrictions, and mainly for the use of local people only. Norway was insisting on social distancing on lifts and restaurants, in China skiers needed a medical health certificate in addition to a quality face mask to ski, and most resorts drastically limited numbers of people on the slopes. Perhaps a sign of things to come?

Sources: Blevins (2020); Griffiths (2020b); Myers (2020); Thorne (2020); Klassen (2020); Strong (2020)

■ References

Alpert, B. & Beilfuss, L. (2020). Millions of layoffs are coming, hotels, airlines, and retailers warn. *Barron's*, 22 March. https://www.barrons.com/articles/millions-of-layoffs-are-coming-hotels-airlines-and-retailers-warn-51584899416

Blevins, J. (2020). Coronavirus forced Colorado's ski industry to shutter. *The Colorado Sun*, 15 April. https://coloradosun.com/2020/04/15/colorado-ski-resorts-shutdown-backstory/

Board of Innovation (2020). *The new low touch economy. How to navigate the world after COVID.* https://www.boardofinnovation.com/low-touch-economy/

Carr, A. & Palmeri, C. (2020). Carnival executives knew they had a virus problem, but kept the party going. *Bloomberg Businessweek*, 16 April. https://www.bloomberg.com/features/2020-carnival-cruise-coronavirus/

Choat, I. & Wilson, A. (2020). How the coronavirus outbreak is affecting travel in Europe. *The Guardian*, 26 February. https://www.theguardian.com/travel/2020/feb/26/how-the-coronavirus-outbreak-is-affecting-travel-in-europe

Clark, D.B. (2020). Inside the Nightmare Voyage of the Diamond Princess. *GQ Magazine*, 30 April. https://www.gq.com/story/inside-diamond-princess-cruise-ship-nightmare-voyage

Cuozzo, S. (2020). NYC's hotel industry bracing for potential coronavirus disaster. *New York Post*, 24 February. https://nypost.com/2020/02/24/nycs-hotel-industry-bracing-for-potential-coronavirus-disaster/

Duke, J. & Bagshaw, E. (2020). Ad campaign to woo 'safety-conscious' tourists away from Asia to Australia amid coronavirus fears. *The Sydney Morning Herald*, 21 February. https://www.smh.com.au/politics/federal/ad-campaign-to-woo-safety-conscious-tourists-away-from-asia-to-australia-amid-coronavirus-fears-20200220-p542n7.html

Eisen, D. (2020). January hotel data highlights coronavirus scourge. *Hospitality Net*, 27 February. https://www.hospitalitynet.org/opinion/4097189.html

Ekstein, N. (2020). Tourism hit from coronavirus will carry into 2021, travel experts say. *Japan Times*, 16 February. https://www.japantimes.co.jp/news/2020/02/16/business/economy-business/tourism-coronavirus-covid19-2021-experts/

Ganguly, S. (2020). Tourists stranded in India because of coronavirus lockdown. *CTV News*, 2 April. https://www.ctvnews.ca/health/coronavirus/tourists-stranded-in-india-because-of-coronavirus-lockdown-1.4878871

Griffiths, J. (2020a). Taiwan's coronavirus response is among the best globally. *CNN.com*, 5 February. https://www.cnn.com/2020/04/04/asia/taiwan-coronavirus-response-who-intl-hnk/index.html

Griffiths, D. (2020b). Chic alpine ski resort became a virus haven while officials dallied. *Bloomberg News*, 11 April. https://www.bloomberg.com/news/articles/2020-04-11/chic-alpine-ski-resort-became-a-virus-haven-as-officials-dallied

Hospitality Net (2020). ITB Berlin canceled. Press Release from *Hospitality Net*, 28 February. https://www.hospitalitynet.org/news/4097246.html

Hudson, S. & Hudson, L. J. (2017). *Marketing for Tourism, Hospitality, and Events*. Sage, London.

Jordan, A.E. (2020). Cruise lines prepare for the future. *The Maritime Executive*, 2 April. https://www.maritime-executive.com/editorials/cruise-lines-prepare-for-the-future

Jozuka, E. (2020). Kyoto launches an 'empty tourism' campaign amid coronavirus outbreak. *CNN*. https://www.msn.com/en-us/travel/news/kyoto-launches-an-empty-tourism-campaign-amid-coronavirus-outbreak/ar-BB106A2h

Klassen, C. (2020). No new COVID cases at Sun Peaks, but resort dealing with economic fallout. *CFJC Today*, 14 April. https://cfjctoday.com/2020/04/14/no-new-COVID-cases-at-sun-peaks-but-resort-dealing-with-economic-fallout/

McCarthy, N. (2020). COVID: Unprecedented decline in air traffic. *Statista*, 31 March. https://www.statista.com/chart/21288/flights-tracked-daily-worldwide/

Miranda, L. & Atkinson, C. (2020). From empty sidewalks to deserted hotels, coronavirus is slamming the tourism industry. *NBC News*, 5 March. https://www.nbcnews.com/business/business-news/empty-sidewalks-deserted-hotels-coronavirus-slamming-tourism-industry-n1147061

Morgan, R. (2020). Coronavirus: Hotel rooms and coach tours canceled in Northern Ireland. *BBC News*, 19 February. https://www.bbc.com/news/uk-northern-ireland-51552556

Myers, P. (2020). France's ski slopes shut down to halt march of coronavirus. *RFI*, 15 March. http://www.rfi.fr/en/france/20200315-france-ski-coronavirus-edouard-philippe-train-bus-sncf-transport

Mzezewa, T. (2020). Alaska tourism officials see an opportunity in coronavirus. *New York Times*, 27 February. https://www.nytimes.com/2020/02/27/travel/coronavirus-travel-alaska.html

National Restaurant Association (2020). New research details early impact of coronavirus pandemic on restaurant industry. *National Restaurant Association*, 25 March. https://www.restaurant.org/Articles/News/Study-details-impact-of-coronavirus-on-restaurants

Nussey, S., & Tajitsu, N. (2020). Japan's olympics delay could cost $6 billion including hit to tourism. Skift, 30 March. https://skift.com/2020/03/30/japans-olympics-delay-could-cost-6-billion-including-hit-to-tourism/

Robbins, T. (2020). Coronavirus: Travel industry bids to keep us booking. *Financial Times*, 5 March. https://www.ft.com/content/c32db48e-5dff-11ea-8033-fa40a0d65a98

Romine, T., Levenson, E. & Morgado, J. (2020). Broadway theaters to suspend all performances because of coronavirus. *CNN*, 12 March. https://edition.cnn.com/2020/03/12/health/broadway-coronavirus-update/index.html

Schultz, K. & Sharma, B. (2020). Stranded abroad, Americans ask: why weren't we warned sooner? *New York Times*, 3 April. https://www.nytimes.com/2020/04/03/world/asia/coronavirus-state-department-tourists.html

Scott, P. (2020). Impact of the coronavirus ripples across Asia: 'It has been quiet, like a cemetery.' *The New York Times*, 23 February. https://www.nytimes.com/2020/02/23/travel/coronavirus-asia-tourism.html

Shiji Distribution Solutions (2020). COVID crisis recovery data from China. https://www.shijigroup.com/insights/china-hospitality-retail-food-covid19-recovery-live

Skopeliti, C. (2020). Jet2 planes turn around in mid-air as firm cancels Spain flights. *The Guardian*, 14 March. https://www.theguardian.com/world/2020/mar/14/jet2-planes-turn-back-in-mid-air-as-firm-cancels-spain-flights-over-coronavirus-measures

STR (2020). STR: US hotel results for week ending 28 March. *Smith Travel Research*. https://str.com/press-release/str-us-hotel-results-week-ending-28-march

Strong, J. (2020). Small businesses lose millions in tourism revenue. *Sun Peaks Independent News*, 6 April. http://sunpeaksnews.com/small-businesses-lose-millions-in-tourism-revenue-34922.htm

The Economic Times (2020). International tourism to plunge up to 30% due to virus: *UNWTO*. https://economictimes.indiatimes.com/news/international/business/international-tourism-to-plunge-up-to-30-due-to-virus-unwto/articleshow/74849024.cms?utm_source=contentofinterest&utm_medium=text&utm_campaign=cppst

Thorne, P. (2020). Ski resorts are re-opening in three countries. In the Snow, 13 April. https://www.inthesnow.com/ski-resorts-are-re-opening-in-three-countries

UNWTO (2020), https://www.unwto.org/news/covid-19-international-tourist-numbers-could-fall-60-80-in-2020

Walsh, Alistair. (2020a). Coronavirus: New Zealand allows stranded tourists to leave. *DW*, 2 April. https://www.dw.com/en/coronavirus-new-zealand-allows-stranded-tourists-to-leave/a-52988164

Walsh, Bryan. (2020b). COVID: The history of pandemics. *BBC Future*, 25 March. https://www.bbc.com/future/article/20200325-COVID-19-the-history-of-pandemics

Yaffe-Bellany, D. (2020). Cruise industry, a symbol of the pandemic, scrambles to survive. *New York Times*, 7 April. https://www.nytimes.com/2020/04/07/business/coronavirus-cruise-. industry-carnival.html

2 Adapting to lockdown

Case study: Richard Branson and the challenges of leadership during the COVID-19 pandemic

"Crisis does not build character, it reveals it" is the old adage, and the COVID-19 outbreak certainly created significant challenges for leaders at all levels and in organizations of all sizes.

Richard Branson – one of the most high-profile leaders in the travel sector - was no exception. He has called the pandemic the most challenging crisis that he has ever faced. In March 2019, your author heard Branson speak at a conference in Salt Lake City, where he was composed, articulate, full of confidence for the future, especially his new venture Virgin Voyages, a cruise line he was gearing up to launch. How things can change in a year!

Figure 2.1: Virgin Atlantic A330 (courtesy of Virgin Atlantic)

Virgin companies employ more than 70,000 workers, across 35 countries, in the travel, hotel and leisure industries, which were among the hardest hit when global travel slowed down in February 2020. In early March, Virgin Voyages organized its maiden voyage but postponed the wider launch of its Scarlet Lady cruise ship; and, shortly after, British airline Flybe collapsed due to weakened demand during the COVID-19 outbreak. Flybe, which once was Europe's largest independent regional carrier, narrowly avoided collapse in January, after being bought by Cyrus Capital, Virgin Atlantic and Stobart. Virgin Galactic, Branson's publicly traded space tourism arm, also saw its shares slump dramatically.

The biggest challenge, he said at the time, is that "there is no money coming in and lots going out". Virgin Atlantic was the first UK airline to seek a bespoke support package from the government early in April, but the airline was told to resubmit its proposal for a £500 million ($620m) coronavirus bailout package after the government was left unimpressed with its initial bid. The money was to come from a split of commercial loans to cover fixed costs over the coming months, such as ticket refunds and airport parking charges. The remainder would come in the form of a credit guarantee that would stop credit card companies from holding back passenger revenues for future bookings from the airline, which had hit its liquidity further. Later that month, Branson attacked the Australian government for refusing to bail out Virgin Australia, after the airline went into voluntary administration due to impact of COVID-19.

The British government took a far tougher line with airlines than the Trump administration had with US carriers. Instead of targeting specific industries for bailouts it put workers before owners and shareholders by guaranteeing all furloughed workers a minimum of 80% of their wages for three months, with an extension if needed. In contrast, US airlines shared a $25 billion bailout, despite the fact that over the last decade the five largest carriers put 96% of the cash they generated into buying back their own shares, pushing up share prices and enriching their investors.

During the crisis, Branson received much negative press in the UK, with an ongoing debate in the media as to whether British tax payers should be bailing out a man who once sued the NHS (National Health Service), and who had not paid income tax in the UK for 14 years. Branson's offer to use his private island in the British Virgin Islands (BVI) as collateral did not seem to strengthen his case. Necker Island and the BVI had often been portrayed as nothing more than a tax haven by some UK politicians and newspapers. Branson had been a resident there since 2006, receiving illustrious guests including Princess Diana and Barack Obama. Branson was also accused of double standards during the crisis, asking staff to take a wage cut for eight weeks. Opposition Labour Party lawmaker Kate Osborne, the second British politician diagnosed with the coronavirus, said Virgin Atlantic's decision was "an absolute disgrace" and Conservative MP Richard Fuller suggested Branson himself pay the wages for his 8,000 plus staff. "Sir Richard Branson's net worth is 3.8 billion pounds ($4.7 billion). If he's able to get 2% interest on that money for eight weeks, he will earn the equivalent of 9.9 million pounds. So I say, Sir Richard Branson, give up your interest on your wealth for eight weeks and pay your employees yourself their unpaid leave."

But Branson fought back against his critics. He committed to injecting $250 million of his own wealth to support his companies, with a large share going to Virgin Atlantic. A Virgin Group spokesperson said "Richard and the Virgin Group have invested hundreds of millions in businesses in the UK, created thousands of jobs and generated significant tax and other revenues to the government for the benefit of the UK economy. Richard now spends most of his time starting not-for-profit ventures and raising millions for charity." Branson had the backing of Airbus, which makes Virgin's planes, and Rolls-Royce, which supplies its jet engines, both warning that if the airline collapsed it could drag them down too.

Branson founded Virgin Atlantic in 1984 and has a 51% stake alongside US airline Delta with 49%. Although the company began as a no-frills cheaper alternative to British Airways, flying a few flights out of London's second airport, Gatwick, it eventually evolved into a full-service competitor pulling in business class passengers with the front of its cabins styled as Upper Class (combining first and business) and a swish Heathrow lounge with its own direct access for those arriving in private limos. However, Virgin Atlantic lost over $100 million in 2017 and 2018, the two most recent years for which its accounts are available.

Some suggested that the problem for Virgin in the midst of the crisis was that it was hard to make a financial case that an international airline could be run like a boutique brand. The scale is too small. The same problem doomed Virgin America. Passengers loved its style and service and it repeatedly won awards that placed it above other domestic carriers, but it couldn't make money and was eventually swallowed by Alaska Airlines, which paid $2.6 billion for it in 2016 and then set about removing all traces of Virgin innovations.

Over the past 30 years, Branson has diversified his Virgin brand into a far-reaching empire, encompassing mobile phone services, rail service, healthcare, cruise ships, hotels and even wedding dresses, as well as his original record label and discount airline. The secret to his expansion lay in licensing the highly regarded Virgin name, a strategy he calls "branded venture capital." Many of his companies have been successful, but others have failed, and these include Virgin Cola, Virgin Vodka, Virgin Clothing, Virgin Cars and Virgin Flowers. In fact, when the author heard Branson speak in 2019, he said his biggest failure was Virgin Cola. Whether or not Virgin Atlantic moves to the top of the failure list, only time will tell. As Branson once said: "As soon as something stops being fun, I think it's time to move on." With the pandemic ravaging tourism and travel across the globe, maybe it's that time?

Sources: Ranosa (2020); Powley, Thomas & Payne (2020); Gill (2020); Hyde (2020) Harper (2020)

Introduction

By mid-April 2020, a third of the global population was under full or partial lockdown. While 'lockdown' was not a technical term used by public-health officials, it referred to anything from mandatory geographic quarantines to non-mandatory recommendations to stay at home, closures of certain types of businesses, or bans on events and gatherings. During this lockdown period, the travel sector worldwide continued to experience a loss of business. For example, Spain's famous annual San Fermin bull-running festival, which usually draws thousands of participants, was canceled because of the coronavirus crisis. "As expected as it was, it still leaves us deeply sad," said acting mayor Ana Elizalde in a statement from the local Pamplona town hall. The July festival, which was made famous in Ernest Hemingway's novel *The Sun Also Rises*, has seldom been canceled in its history. Other major European tourist events were canceled, including Oktoberfest, the famous annual German beer-drinking festival which traditionally sees six million people travel to Munich.

Unfortunately, as the coronavirus spread across the world, so too did misinformation about it, despite an aggressive effort by social media companies to prevent its dissemination. Facebook, Google and Twitter were removing false information about the coronavirus, such as miracle cures, secret labs and government plots, as fast as they could find it, and worked with the World Health Organization (WHO) and other government organizations to ensure that people got accurate information (Frenkel, Alba & Zhong, 2020). One report from the Reboot Foundation

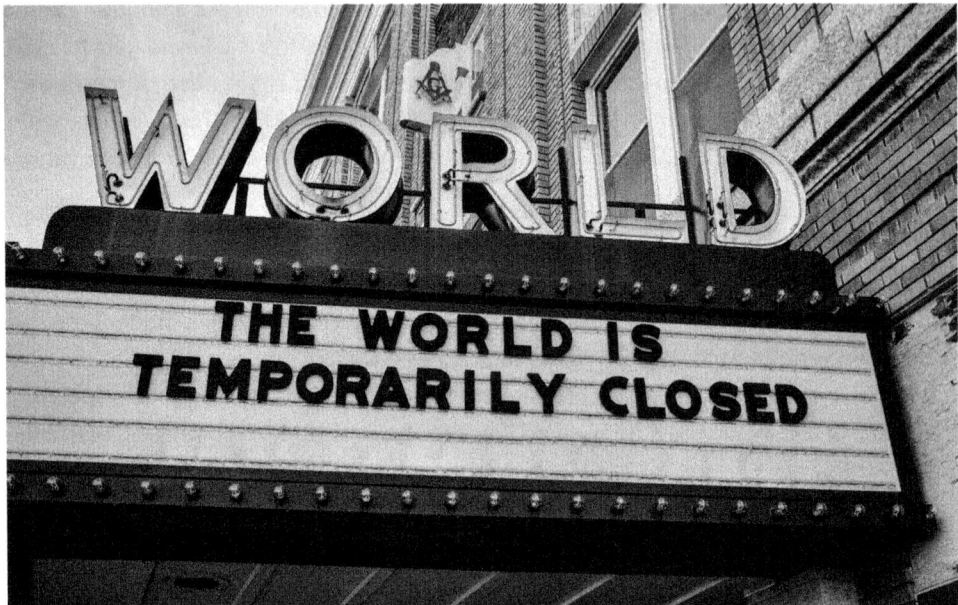

Figure 2.2:. By mid-April 2020, a third of the global population was under full or partial lockdown (photo by Edwin Hooper on Unsplash)

found that, not only was social media rife with misinformation on COVID-19, but the more time people spent on platforms like Twitter, the less informed they were on the spread and prevention of the virus (Bouygues, 2020).

Almost 100 years previously, during the deadliest pandemic in modern history, there was the opposite problem. A lack of news about the Spanish Flu exacerbated the issue. Wartime censors suppressed news of the flu – which eventually killed between 40 and 50 million – to avoid affecting morale. In fact, it only became known as 'Spanish Flu' because Spain remained neutral during World War I, and so the Spanish media was free to report on it in gory detail. Since nations undergoing a media blackout could only read in depth accounts from Spanish news sources, they naturally assumed that the country was the pandemic's ground zero. Scientists are still unsure of its source, but it is unlikely that the Spanish Flu originated in Spain at all (Andrews, 2016).

Leadership during the crisis

According to historian and author Yuval Noah Harari (2020), the COVID-19 pandemic exposed a lack of global leadership. He said at the time "humanity faces an acute crisis today not only due to the coronavirus, but also because of a lack of trust between humans". He suggests that to defeat any epidemic, people need to trust scientific experts, citizens need to trust public authorities, and countries need to trust each other. "Over the last few years, irresponsible politicians have deliberately undermined trust in science, in public authorities and in international cooperation. As a result, we are now facing this crisis bereft of global leaders that can inspire, organize and finance a coordinated global response."

Figure 2.3: Social distancing became a common practice during lockdown – as well as a new term in the Oxford English Dictionary (photo by Jordan Hopkins on Unsplash)

For leaders though, there was no easy route through the crisis – as we saw from the case study on Richard Branson. Besides the obvious problem of selecting the correct strategy to take, leaders faced the monumental task of reassuring the public and persuading them to follow through on government decisions – even when measures such as social distancing with its devastating knock-on effect on employment – come at great personal cost (Robson, 2020). Many observers made the point that all the countries that had a commendable response to COVID-19 had one thing in common – female leadership. Countries like New Zealand, Germany, Finland and Taiwan, for example – all with women leaders – were applauded for their empathetic and transparent approach, with due consideration given to looking after their citizens' social and economic welfare.

Not all female leaders made the headlines for the right reasons. Towards the end of April, in an interview with CNN's Anderson Cooper, Las Vegas Mayor Carolyn Goodman said she was willing to offer the city as a 'control group' if Las Vegas was allowed to reopen. Goodman had been an adamant proponent of reopening the city, even calling the shutdowns 'total insanity'. The interview quickly began trending on social media, with almost everyone appearing to react negatively, notably Nevada Governor Steve Sisolak who quashed the idea. Late-night television host and former Las Vegas resident Jimmy Kimmel called Goodman an 'embarrassment to my hometown'.

As Boin et al. (2005) have said, it is often the content of the leader's messaging that ultimately determines the public's trust – not simply 'doing the right thing' on the ground. In their book on crisis management, they say that effective leaders must excel in the communication dimension of crisis management. "If they do not get their message to the public about the causes, consequences and cures of the crisis, others will," they say. "The political communications process is highly competitive: each and every detail of words, pictures, gestures, and performance matters" (pp. 70). The communications efforts of those in the tourism and hospitality sector during the crisis will be looked at in more detail in Chapter 3.

Were tourism and hospitality leaders ready for this pandemic? Beirman (2003) has suggested that the first step to the management of any crisis is to establish a crisis management team *before* any crisis breaks out. Such a team should be assigned key roles, such as media and public relations, relations with the travel industry in source markets, and destination response co-ordination with the local tourism industry. Disney, for example, has a Global Crisis Management team whose mission is to "collaborate with key business partners and customers to support all Disney (and its affiliates) employees in building, sustaining and improving Disney's capability to prepare for, respond to and recover from all hazards". Part of Disney's crisis management response during the COVID-19 pandemic was to persuade CEO Bob Iger to postpone semi-retirement and retake control of the entertainment group as it braced for changes caused by the corona-virus. Iger's 15-year stint as chief executive was due to come to an end, with Bob Chapek, the head of Disney's theme park business, unveiled as his successor in

February. But Iger, who had already said he would not take his $3 million salary during the pandemic, decided to remain at the helm to help steer Disney through a period of unprecedented turmoil that was reportedly costing the company $30 million a day (Davies, 2020).

In theory, the middle of a crisis is not a good time to formulate a strategy for dealing with it, but for many organizations that is precisely when the conversation begins. One study in Canada in 2019 found that only 40% of businesses had a reputation recovery plan included in their overarching crisis communications strategy, even though roughly 60% said that damage to a company's reputation is the most difficult aspect of a crisis from which to recover. "Companies that wait until a crisis strikes to prepare a recovery strategy are often too late to avoid significant and lasting damage to their reputation and bottom line," said Wojtek Dabrowski, Founder and Managing Partner of Provident Communications Inc. who conducted the study (Provident Communications, 2019).

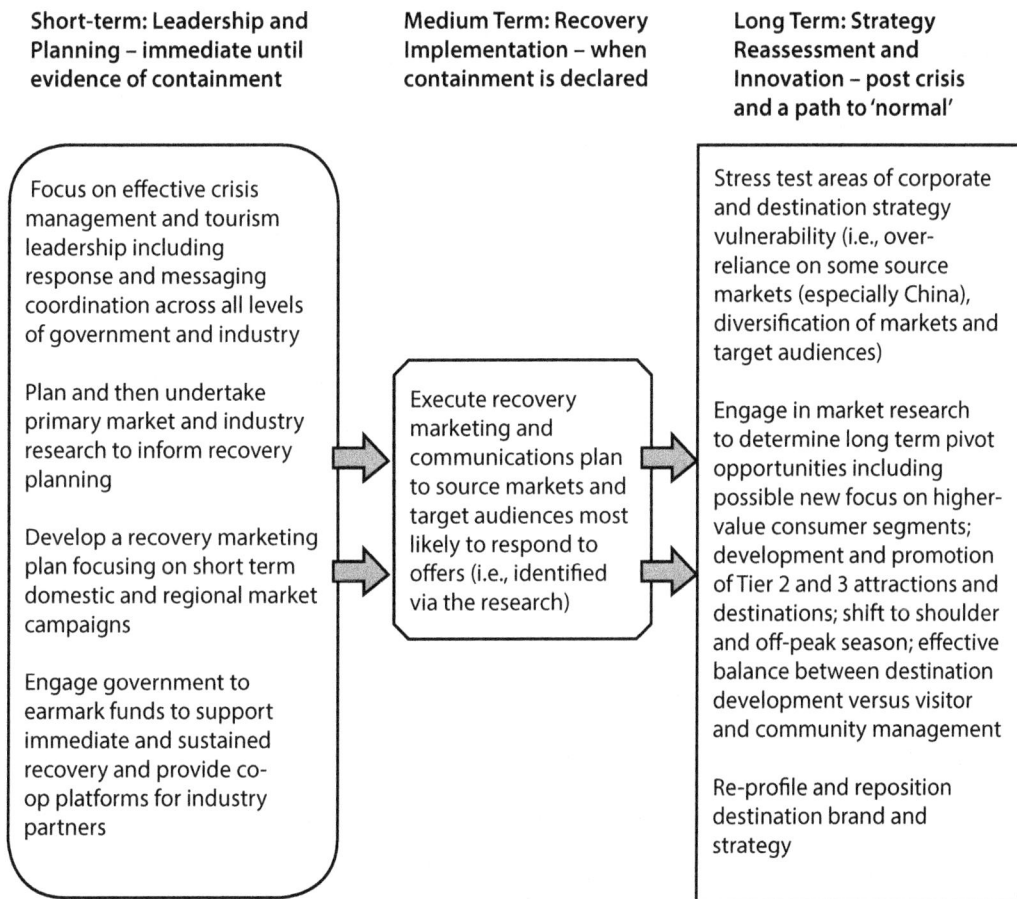

Short-term: Leadership and Planning – immediate until evidence of containment	Medium Term: Recovery Implementation – when containment is declared	Long Term: Strategy Reassessment and Innovation – post crisis and a path to 'normal'

Focus on effective crisis management and tourism leadership including response and messaging coordination across all levels of government and industry

Plan and then undertake primary market and industry research to inform recovery planning

Develop a recovery marketing plan focusing on short term domestic and regional market campaigns

Engage government to earmark funds to support immediate and sustained recovery and provide co-op platforms for industry partners

Execute recovery marketing and communications plan to source markets and target audiences most likely to respond to offers (i.e., identified via the research)

Stress test areas of corporate and destination strategy vulnerability (i.e., over-reliance on some source markets (especially China), diversification of markets and target audiences)

Engage in market research to determine long term pivot opportunities including possible new focus on higher-value consumer segments; development and promotion of Tier 2 and 3 attractions and destinations; shift to shoulder and off-peak season; effective balance between destination development versus visitor and community management

Re-profile and reposition destination brand and strategy

Figure 2.4: Recommended strategies for DMOs during the COVID-19 outbreak (based on Twenty31, 2020)

Tourism consultants Twenty31 suggested that the response from the travel trade at both a public and private sector level was disjointed. They believed leaders needed to move beyond damage control and market reassurance towards developing a post-crisis strategy. Figure 2.4 summarizes their recommended strategies for tourism destinations during the COVID-19 outbreak.

Staying afloat

During this lockdown period, the United Nations World Tourism Organization (UNWTO) provided guidance and support for the global tourism sector. As the leading international organization in the field of tourism, UNWTO promotes tourism as a driver of economic growth, inclusive development and environmental sustainability, and offers leadership and support to the sector in advancing knowledge and tourism policies worldwide. UNWTO's membership includes 159 countries, six Associate Members and over 500 Affiliate Members representing the private sector, educational institutions, tourism associations and local tourism authorities. From the start of the outbreak, UNWTO worked closely with the WHO to ensure public health measures were implemented in ways that minimized any unnecessary disruption to travel and trade. UNWTO called for financial and political support for recovery measures aimed at tourism. UNWTO Secretary-General Zurab Pololikashvili stressed that "small and medium sized enterprises make up around 80% of the tourism sector and are particularly exposed with millions of livelihoods across the world, including within vulnerable communities, relying on tourism" (Hospitality Net, 2020a).

One of the initiatives the UNWTO launched to support the industry in collaboration with WHO, was the "Healing Solutions for Tourism Challenge". This was a global call for entrepreneurs and innovators asking them to submit ideas that could help the tourism sector mitigate the impact of the pandemic and kick-start recovery efforts. In particular, the challenge was aimed at finding ideas that could make a difference right away: for destinations, for businesses and for public health efforts. The winners of the challenge were invited to pitch their ideas to representatives of more than 150 governments. UNWTO Secretary-General Zurab Pololikashvili explained: "Tourism is the sector that has been hit the hardest by COVID-19. Our response needs to be strong and united. We also need to embrace innovation. I call on all entrepreneurs and innovators with ideas that are developed and ready to be put into action to share them with us. In particular, we want to hear ideas that will help communities recover from this crisis, economically and socially, as well as ideas that can contribute to the public health response."

UNWTO received over 1,000 applications from over 100 countries for the challenge.

Figure 2.5: UNWTO Healing Solutions for Tourism Challenge

Governments also stepped in to support the industry. In the US at the end of March, President Trump signed the CARES Act (Coronavirus Aid, Relief and Economic Security Act), a $2 trillion stimulus meant to keep the US economic engine running. Certain sectors of the travel industry benefited more from this act than others. Airlines, for example, had access to roughly $50 billion, half in grants and half in loans. After accepting federal money, the Act barred them from laying off or furloughing frontline employees – typically flight attendants, pilots, gate and airport agents – through to September. In addition, the Act limited airlines from buying back shares or increasing executive compensation. Airports, who were seeing a collapse in airport revenues – with far less money coming in from landing fees, parking, and concessions than expected – also received $10 billion from the government.

Individual hotel owners and big chains got a boost from the CARES Act. Hotel operators were included in the group eligible to tap into federal small business loans set aside for businesses negatively impacted by coronavirus. A provision was added to enable many one-off hotel operators to qualify for small business benefits even if they operated under the flag of a larger brand like Marriott or Hilton. The larger brands were not eligible for these federal loans. "We have been advocating on behalf of these small business owners for the emergency loans that will allow them to remain solvent while operations are suspended at their hotels," a Hilton representative said (Sumers, 2020). Larger hoteliers were not entirely ignored by Washington. The new law's Coronavirus Economic Stabilization Act of 2020 subset offered $454 billion in liquidity to affected industries like the hospitality sector. The measure was meant to keep workers on payroll at mid-sized companies with between 500 and 10,000 employees. To be eligible, a larger hotel company would have to use funds to support business operations and keep 90% of its workforce with full salaries and benefits through the end of June.

There was some backlash over the distribution of the funds, however. Several media outlets revealed how large chunks of the package were taken up by chain restaurants, hoteliers and publicly traded corporations, rather than small, local businesses. This prompted some chains to return the loans to the government. Burger chain Shake Shack, for example, returned a $10 million loan it received from the US government. CEO Randy Garutti and chairman Danny Meyer revealed their decision to give back the funding in an open letter, saying that the NYSE-listed company no longer needed the money because they were "fortunate to have access to capital that others do not" (Toh, 2020). The company expected to be able to raise up to $75 million from investors by selling shares.

Interestingly, the cruise industry was excluded from the stimulus package. The bill stated that a business eligible for the allocated $500 billion in government loans and loan guarantees must have been "created or organized in the US or under the laws of the US" and have "significant operations in and a majority of its employees based in the US". None of the big three cruise lines – Carnival Corp, Royal Caribbean, and Norwegian Cruise Lines – could make that claim. Each has physical headquarters in Miami, but all are incorporated and flag their ships in other nations. A large proportion of their workforce comes from other countries, and they pay comparatively little federal income tax. They also do not have to abide by US labor laws.

Other countries introduced schemes to help tourism and hospitality businesses support staff and prevent a slew of redundancies. In the UK, for example, the government's Coronavirus Job Retention scheme lessened the burden of keeping staff on the payroll during the lockdown period (Ogden, 2020). The scheme enabled businesses to register employees who could no longer work as 'furloughed workers' through an online portal, which then required the Government to reimburse up to 80% of these wages (which was capped at £2,500 per month). The government also announced that all hospitality, retail and leisure businesses across the UK would not have to pay business rates for the 2020/21 tax year.

In France, a hardship grant of approximately $1,600 a month, as well as suspension of taxes and social security contributions, was available for small business owners. The package of assistance was put in place for a minimum of three months, with provision for renewal if necessary. Specific aid was granted to businesses in the tourism, culture, hospitality and events sectors. France welcomed 92.9 million foreign visitors in 2019, and yet tourism was virtually at a standstill during the COVID-19 lockdown. Most neighboring countries had closed their borders, and Air France, the flag carrier, and EasyJet, the second biggest airline in the country, grounded 90% of their fleets.

Figure 2.6: French tourism was almost at a standstill during the COVID-19 lockdown (photo by Fran Boloni on Unsplash)

But these stimulus packages did not prevent some travel companies from going under. In the airline sector, first Virgin Australia went into voluntary administration after the Australian government rejected the airline's plea for a loan. Then four subsidiaries of Scandinavian budget carrier Norwegian filed for bankruptcy. The companies employed more than 4,000 pilots and cabin crew members in Denmark and Sweden. Norway-based employees' salaries were covered by the country's government while they are furloughed, but this was not true for crewmembers based in Denmark and Sweden. Elsewhere, Air Mauritius entered into voluntary administration, following "a complete erosion of the Company's revenue base" as a result of the coronavirus pandemic, and South African Airways was under threat of liquidation.

Adapting to lockdown

For the travel industry, the lockdown period was a time of adaptation. In the airline sector, some grounded their entire fleets, while others kept a few key aircraft ready to perform repatriation flights, or fly critical medical supplies and other cargo around the world. But most planes that would ordinarily be in the air were on tarmac, and on top of the multitude of economic and logistical problems this posed for airlines, employees and passengers, storage of these planes became a challenge (Walton, 2020). There wasn't enough space at most global hub airports to park all the planes that are notionally based there. Some, therefore, used dedicated storage facilities. Southwest and Delta Air Lines, for example, stored aircraft at Victorville, a former US air base in California that now serves as a logistics hub for business, military and freight aviation. Victorville is also a 'desert boneyard' famous for its rows and rows of mothballed aircraft, as is Roswell, in the US state of New Mexico, which also became a temporary home for Boeing's 737, 757, 767 and 777 models. The airport used to be a major US Air Force base during the Cold War and had space to spare after the base's closure.

Since Europe doesn't have the space or ultra-dry desert climates for boneyards, airlines there were parking their aircraft at a variety of different airports. Lufthansa, for example, parked most of its planes at airports in Frankfurt, Munich and Berlin. Parked aircraft need to be checked regularly, usually once a week, by qualified mechanics to ensure that there is no damage, and it takes some 60 working hours to bring the average plane back into service. This meant that relaunching the global commercial air fleet would have to be a staggered process, rather than an immediate spike back to the way things were – and some planes would not return at all (Walton, 2020).

Hotels were having to adapt to very low occupancy rates during the lockdown period, and many closed their doors. Early in April, Accor Hotels announced that more than half of its branded hotels worldwide were currently closed and more than two-thirds would likely be closed in the coming weeks. Accor announced travel bans, hiring freezes, reduced schedules, and/or furloughing for 75% of its

Figure 2.7: Many aircraft were stored at desert boneyards during the lockdown (photo by Jim Bigham)

global head office teams, in addition to other cost reduction efforts, extending to sales, marketing, and IT. Choice Hotels International also announced several mitigation efforts, including reduced compensation for the Board of Directors, CEO, and other executive officers for the remainder of 2020. Montage International focused on the well-being of employees and rolled out several initiatives to support associates across all properties in their Montage and Pendry portfolio. The program included the Hearts of Montage Relief Fund, which guests could donate to directly or purchase a $500 Montage certificate for a one-night stay. In addition, the company sponsored the distribution of meals and care packages for families at all Montage and Pendry properties twice a month. Hyatt similarly launched a Hyatt Care Fund to help employees across its owned, managed, and franchised hotels and corporate offices who had been financially impacted by COVID-19 following furloughs or reduced hours.

Wyndham took a different, more creative and resourceful approach to supporting staff by partnering with leading organizations to redistribute the workforce. Hotel-level personnel and corporate team members, whose positions had been furloughed or eliminated, were given access to available full-time, part-time, or temporary positions in other sectors that were staying open during the crisis including retail, grocery, and senior living. The companies Wyndham partnered with included Albertsons Companies, Amazon, Dollar Tree & Family Dollar, Domino's, Lowe's, Pizza Hut, Senior Living Works, University of Texas Medical Branch, Walgreens, and Walmart. Hilton also introduced a tool called the Hilton Workforce Resource Center that connected furloughed workers with short-term jobs created by the crisis.

Some hotels started offering daily room rates to guests who were looking to escape lockdown at home and get some work done. Drury Hotels, for example, invited customers to 'Find Your Happy Space', offering 'peace, quiet and Wi-Fi for only $39 a day'. Guests could check in between 7am and 6pm. Others started to change operations to emphasize cleanliness and safety. The Westin Houston Medical Center became the first hotel in the country to use robots to combat germs (Rosen, 2020). The robots use UV light to kill the germs, even disinfecting products such as bath amenities and coffee, which are then stored in sealed bags for guests. All public areas and guest rooms were also sanitized using a hospital-grade disinfectant, and hotel staff removed all nonessential items such as magazines and decorative pillows. The hotel employed two of the germ-killing robots on the property to clean its 273 rooms. The robots clean every guest room between stays, as well as common spaces such as the lobby, restaurant, café, bar and meeting room daily. These enhanced disinfectant protocols have been used in over 500 hospitals around the world.

Casinos were also having to adapt and look for alternative revenue streams during this lockdown period. Before COVID-19, casinos in Las Vegas would have laughed at the prospect of an e sports-centric betting operation (Reames, 2020). But a month after traditional sports across the US shut down due to the virus, the

Nevada Gaming Control Board (GCB) approved bets on four different esports series, adding to the slowly growing betting options for competitive gaming. As the regulatory government body that oversees America's gambling capital, the Nevada GCB serves as a barometer for much of the sports betting world. After first approving bets on Counter-Strike: Global Offensive's ESL Pro League Season 11 in North America and Germany, the board then opened up an ESL Dota 2 event and eNASCAR. Then the biggest esports leagues in the US followed – the League of Legends Championship Series in North America; the League of Legends European Championship and Overwatch League; and the Call of Duty League.

In the restaurant sector, while most establishments were not allowed to offer dine-in services, some made extra efforts to promote their takeout or delivery service. For many this was the only way to prevent them from closing down. Even before lockdown, the delivery business had been increasing, with restaurants partnering with one or more of the four dominant food delivery companies, namely UberEats, Grubhub, Postmates, and DoorDash, for food delivery. McDonald's, for example, added DoorDash as a new partner for food delivery in July 2019. Before that, McDonald's had an exclusive partnership with UberEats, which serves about 64% of McDonald's US stores. Delivery makes up about 2-3% of the chain's business, totally about $3 billion (Kwok, 2020). During the crisis, this was heavily advertised as 'McDelivery'.

In Canada, Wednesdays became the 'official' takeout day. Canada Takeout, a group supported by hundreds of restaurant owners and workers across the country, encouraged Canadians to order a takeout meal every Wednesday to help support restaurants that had to close their doors. To kick off the #TakeoutDay campaign, a variety show took place on April 22 on Facebook live. It featured musical entertainment from Canada's top talent and celebrated athletes. Friends and family could enjoy dinner and a concert together virtually and do their part to help restaurants on #TakeoutDay. "Restaurants are vital to the social and economic fabric of communities across Canada, but operating a foodservice business is tough, even in the best of times," said Shanna Munro, Restaurants Canada President and CEO. "Not only was our industry among the first to feel the impacts of COVID-19, we've been one of the hardest hit so far, with nearly two thirds of our workforce now lost. We would love to see Canadians embrace #TakeoutDay and support those restaurants who are still able to operate through takeout or delivery" (Canada Takeout, 2020).

For some workers in the travel industry, cancelations and closures meant that they were busier than ever. Travel agencies and brokers, for example, were suffering financially, but many were still working around the clock for their clients. "Travel agents have probably never been as busy as they are now, dealing with client refunds and queries," said Andrew Olsen, chief executive of Travel Agents' Association New Zealand (Kenny, 2020). Olsen thought that, as with many service sector and hospitality businesses, travel agencies would not survive COVID-19 and the consequent economic recession. "I can't think of another

2

Figure 2.8: Restaurants around the world promoted takeout or delivery after being forced to close (photo by Kseniia Ilinykh on Unsplash)

Figure 2.9: Canada Takeout Instagram post to promote #TakeoutDay (courtesy of Canada Takeout)

industry where everyone is handing back everything they made." Steve Lee of New Zealand Travel Brokers said the coronavirus pandemic would probably result in more travel agencies turning to remote working to minimize overhead costs. "Our guys have worked tirelessly on behalf of their clients. We said to them at the very beginning of this thing, this was their opportunity to build client relationships for life. If anything good can come out of this, it's strengthened client relationships" (Kenny, 2020).

Tour operators were also busy dealing with cancelation. TUI, the world's largest tour operator, had to cancel thousands of holidays, and had trouble keeping up with demands for refunds or alternative packages. All travel companies in Europe were required to pay a full cash refund with two weeks of canceling a trip, but this deadline was a huge challenge for travel companies across the board. TUI managed to secure a $2 billion bridge loan from the German policy bank KfW in order to reduce the impact of the COVID-19 crisis, and reassured investors and customers that they were in good health financially. "TUI is a very healthy company. We were economically successful before the crisis and will be again after the crisis. Our business model is intact and we have over 21 million loyal customers. However, we are currently facing unprecedented international travel restrictions. As a result, we are temporarily a company with no product and no revenue," said TUI CEO Fritz Joussen in a statement. "The commitment of the KfW bridge loan is an important first step for TUI to successfully bridge the current exceptional situation" (Maritime Executive, 2020).

Figure 2.10: The Sundance Car Café (courtesy of Sundance Resort)

The lockdown did inspire some destinations to think creatively about how to offer a unique hospitality experience in a safe environment. Sundance Resort in Utah (owned incidentally by Robert Redford) introduced the Sundance Car Café in late April. Although most of the resort was closed, they invited guests to "join us for a delicious meal and spectacular views" whereby visitors could make a reservation online, drive their car to Sundance where they would be directed to a parking/dining spot with a spectacular view, and the meal would be brought directly to the vehicle. The Sundance Car Café served up guest favorites from three resort restaurants. Some entrepreneurial restaurants, meanwhile, turned their parking lots into drive-in movie theatres. Peoples Restaurant and Lounge in Corpus Christi, Texas, for example, embraced this new way of serving its customers. "We have converted our restaurant into a Drive-in Movie theater showing older and newer classics while at the same time providing excellent food and service," states its website. "We will continue this until we feel it is safe for the community and us to return to our table side service" (Gillespie, 2020).

Some destinations saw the lockdown as an opportunity to allow consumers to experience their products and experiences virtually or remotely. Destination marketing using virtual reality (VR) was already gathering momentum before the COVID-19 outbreak. VR is a computer-generated simulation of an environment that can be interacted with in a seemingly real or physical way by a person using special electronic equipment, such as a helmet with a screen inside or gloves fitted with sensors. British Columbia in Canada was one of the first destinations to use VR for tourism marketing, providing trade, media partners and end consumers with a unique way to experience the province from their desk chairs. Marriott Hotels has also been testing virtual reality with the 'teleporter', a telephone booth-like structure equipped with a headset, wireless headphone and 4-D sensory elements to provide a virtual travel experience.

During the lockdown, the Faroe Islands launched a new remote tourism initiative whereby curious tourists could take a hike, go kayaking, or jump in a helicopter in the Faroes from their sofa. Through a web interface, people from anywhere in the world could take control of a Faroese person equipped with a live camera and microphone in minute-long blocks, telling them where to go and what to look at. "When the travel bans began to escalate, we wondered how we could recreate a Faroe Islands' experience for those who had to cancel or postpone their trip to the Faroe Islands and for everyone else stuck at home," said Guðrið Højgaard, director of Visit Faroe Islands, in a statement. "The result is this new platform to enable those in isolation to take a walk across our wild landscapes, to regain a sense of freedom and to explore beyond their own four walls" (Collins, 2020). The Faroe Islands has a history of creative tourism initiatives, including strapping cameras to sheep and uploading the footage to Google, and closing the islands once a year to everyone except for volunteers who help with island maintenance.

Figure 2.11: Faroe Islands remote tourism campaign (courtesy of Visit Faroe Islands)

Philanthropy

Although the tourism and hospitality industry may have been one of the worst hit sectors of the economy, it was also an industry that stepped up to the table to assist those in need. It was mentioned previously that a number of airlines were called on to repatriate stranded travelers, and some airlines refitted planes to fly essential medical supplies and other cargo around the world. Air Canada, for example, reconfigured the cabins of three of its Boeing 777-300ER aircraft to give them additional cargo capacity. "Bringing critical medical and other vital supplies rapidly to Canada and helping distribute them across the country is imperative to combating the COVID-19 crisis. The transformation of the Boeing 777-300ERs, our largest international wide-body aircraft, doubles the capacity of each flight and will enable more goods to move more quickly," said Tim Strauss, Vice President of Cargo at Air Canada (Air Canada, 2020).

The accommodations sector was also able to offer a helping hand. At the end of March, Airbnb announced a global initiative to help connect those responding to the COVID-19 pandemic with safe and convenient places to stay while they carried out their critical work. The company's goal was to help house 100,000 healthcare professionals, relief workers, and first responders around the world. Airbnb waived all fees for stays arranged through this initiative. "Medical workers and first responders are providing lifesaving support during the coronavirus outbreak and we want to help," said Airbnb's Co-founder Joe Gebbia. "We've heard from countless hosts around the world who want to provide a comforting home to heroic first responders. We are connecting our nonprofit partners, government agencies and others with our incredible host community to work together in these extraordinary times" (Hospitality Net, 2020b). The initiative built on two pilot programs in Italy and France where doctors, nurses, caregivers and other medical support staff who were responding to the outbreak could access free accommodation through Airbnb. Hosts could opt into the program and had the option of opening their homes for free through Airbnb's Open Homes platform, created in 2012 to meet the needs of people requiring emergency housing. If hosts were not able to host for free, Airbnb would still waive all fees on the stay.

The Walt Disney Company also came up with some creative ways to support the fight against COVID-19 (and generate a little appropriate publicity along the way!) Disney Parks gave 150,000 rain ponchos to MedShare, a humanitarian aid organization, for distribution to hospitals in need. This donation was inspired by nurses across the US, who inventively utilized the rain ponchos to protect their clothing and prolong the use of personal protective equipment (PPE), while also freeing up needed gowns. Disney Parks, Experiences and Products also donated more than 100,000 N95 masks to the states of New York, California and Florida. Then, when Walt Disney Television's productions were forced to shut down, their Los Angeles-based productions *Grey's Anatomy* and *Station 19* sent unused gloves, gowns, masks and other medical supplies to Los Angeles County + USC Medical Center. Other TV series that were halting production sent food donations to area food banks in California, Georgia, New York and Canada.

Finally, US fast-food chain Chick-fil-A chipped in with a relief effort dedicated to helping local franchisees continue their support of those impacted by COVID-19 in the communities they serve. Anchored by a $10.8 million fund that Chick-fil-A distributed to local communities through its network of more than 1,800 independent owner-operators, funds were available through June 2020. These were intended to have an immediate impact on the needs of local communities, including food donations or items for first responders, health care workers, Chick-fil-A Team Members and their families. "Striving to be a supportive, caring and generous neighbor is in our DNA. Our restaurant Operators give back locally in so many ways, and this time is no exception," said Chick-fil-A Chairman and Chief Executive Officer Dan Cathy (QSR, 2020). Chick-fil-A, Inc. was also helping to feed those in need during this time. More than 5,000 meals were donated to school systems, healthcare workers and others on the front lines in the restaurant company's hometown of Atlanta, Georgia.

Case study: Hotels pivot to lend a helping hand

Figure 2.12: Hotel Revival, Baltimore (courtesy of the Hotel Revival)

The hotel industry worldwide stepped up and answered the call of public health to support the pandemic by providing overflow capacity for hospitals, or offering people a facility to self-isolate and protect their families. In the US, the American Hotel and Lodging Association (AHLA) launched 'Hospitality for Hope', an initiative that identified hotel properties that had volunteered to provide temporary housing for emergency and healthcare workers. Participating hotels could potentially be used as emergency hospitals or housing for those who required quarantine. Over 16,000 hotels had signed up by mid-April. Hospitality for Hope was created with the idea to boost collaboration between the hotel industry and local, state, and federal governments to provide much needed assistance during the crisis. Government officials were able to search properties in a special AHLA database based on geographic location. "As an industry of people taking care of people, the hotel industry is uniquely positioned to support and help strengthen our communities and first responders who are on the frontlines of dealing with this ongoing public health crisis," said Chip Rogers, president and CEO of the AHLA. "Hotels have always been an active member of our local communities, and this time is no different."

Many of the big hotel brands provided assistance. Hilton, in partnership with American Express, donated thousands of hotel rooms to doctors, nurses, and other medical professionals, in response to the novel coronavirus pandemic, partnering with professional organizations of medical workers and first responders to help connect individuals in need with rooms around the country. William Jaquis, MD, FACEP, president of the American College of Emergency Physicians, said the room donation was a welcome relief for the thousands of medical staff enduring long hours under challenging circumstances: "Knowing that there is a safe, clean and comfortable hotel room waiting for you at the end of a long shift can make all the difference in the world right now." Hilton said participating hotels would be staffed by team members who had received additional training on relevant health and safety measures to safeguard their own and their guests' well-being. Hotel rooms and common areas would all be sanitized using industrial-grade cleaners and updated cleaning protocols.

Marriott launched a similar program called Rooms for Responders, in collaboration with its credit card partners, American Express and JPMorgan Chase, to provide up to $10 million worth of hotel stays at no cost to frontline healthcare workers in select cities. Administered by the Emergency Nurses Association (ENA) and the American College of Emergency Physicians (ACEP), the program matched frontline nurses and doctors with approximately 100,000 room nights at participating hotels in New York City, Newark, Washington, D.C., Chicago, Los Angeles, Detroit, Las Vegas, and New Orleans. Marriott Bonvoy members could also donate points to relief organizations, including the American Red Cross, International Federation of Red Cross and Red Crescent Societies, UNICEF, and World Central Kitchen. Additionally, through its Community Caregiver Program, Marriot offered special rates for first responders and healthcare professionals who wanted to book rooms near hospitals.

In France, Accor Hotels opened up properties to frontline medical staff, and the French people fighting against the spread of COVID-19. The company also offered 2,000 hotel beds to homeless people throughout France. In collaboration with AP-HP university hospitality trust and its partners, Accor housed medical staff by offering accommodations near their jobs through the Coronavirus Emergency Desk Accor (CEDA) platform. Requests from public authorities and professional associations were also centralized on the CEDA platform.

In Mumbai, the Indian Hotels Company Limited (IHCL), the hospitality brand of the Tata Group, provided free rooms to medical personnel who were treating coronavirus patients. Rooms were available across five hotels in Mumbai, namely Taj Mahal Palace, Taj Lands End, Taj Santacruz, The President and Ginger MIDC Andheri. "We deeply value the contribution from the medical community and will continue to work with them as well as the local authorities as we navigate through this crisis," an IHCL spokesperson said. The move came after reports surfaced of medical professionals and other frontline workers facing social stigma in many cities. As the news went viral, people lauded the group and group founder Ratan Tata for going above and beyond to help those fighting against the infectious disease. Other hotel chains in India followed suit. The Lalit, Lemon Tree, Radisson Hotels, InterContinental Hotels Group (IHG) and many others set aside rooms for quarantining patients, or for medical staff battling the COVID-19 outbreak. Some Indian hotels also provided meals for those at the frontline. "Going by estimates and inputs received from members in different parts of India, in total around 45,000 rooms have been offered to the local authorities and NGOs," said Gurbaxish Singh Kohli, vice president of The Federation of Hotel and Restaurant Associations of India. "The hospitality industry and its entrepreneurs have opened their doors to support the people and the central, state and local authorities to combat the COVID-19 pandemic."

Smaller hotels also helped out. The Hotel Revival, a boutique 107-room hotel in Baltimore, opened its doors as a community resource and anchor during the COVID-19 outbreak, offering free rooms, rent-free spaces for local food suppliers, and bagged lunches for medical staff. Police officers, fire fighters and military personnel could check in for free stays at the hotels, while rooms were offered to doctors and nurses at a discounted rate. "We want to care for first responders, healthcare professionals, small businesses, and those who've been impacted by these hard times, as we are all in this together," said Donte Johnson, general manager of Hotel Revival. "It's our goal to emerge with strength on the other side of this crisis – and to support our community along the way." The hotel hosted weekly lunch and produce distributions for the community at large from the end of March. These lunches and bags of produce were offered to anyone in need, which included medical staff. The hotel donated the excess lunches and produce to local senior centers and businesses in need. The hotel has always placed a major focus on the community, which is why they were so quick to respond as things spiraled downhill as a result of the virus.

Sources: Rizzo (2020); Brady (2020); Chaturvedi (2020); Fox (2020).

■   References

Air Canada (2020). Air Canada reconfigures passenger cabins on three aircraft to transport more vital supplies and necessary cargo. *Air Canada Press Release*, 11 April. https://aircanada.mediaroom.com/2020-04-11-Air-Canada-Reconfigures-Passenger-Cabins-on-Three-Aircraft-to-Transport-More-Vital-Supplies-and-Necessary-Cargo

Andrews, E. (2020). Why was it called the 'Spanish Flu?' *History*, 27 March. https://www.history.com/news/why-was-it-called-the-spanish-flu

Beirman, D. (2003) *Restoring Destinations in Crisis: A Strategic Marketing Approach.* Wallingford: CABI International.

Boin, A., 't Hart, P., Stern, E. & Sundelius, B. (2005). *The Politics of Crisis Management: Public Leadership Under Pressure.* Cambridge: Cambridge University Press.

Bouygues, H.L. (2020). Going viral: How social media is making the spread of coronavirus worse. *Forbes*, 2 April. https://www.forbes.com/sites/

helenleebouygues/2020/04/02/going-viral-how-social-media-is-making-the-spread-of-coronavirus-worse/#f1b116941b29

Brady, P. (2020). Hilton and American Express team up to donate 1 million hotel rooms to coronavirus first responders. *Travel + Leisure*, 21 April. https://www.travelandleisure.com/hotels-resorts/hilton-and-american-e...ess-team-up-to-donate-1-million-rooms-to-coronavirus-first-responders

Canada Takeout (2020). Canadians rally to save restaurants with #TakeoutDay. Press Release, 8 April. https://canadatakeout.com/media/

Chaturvedi, A. (2020). Hotels open doors in fight against Coronavirus. *Economic Times*, 4 April. https://economictimes.indiatimes.com/industry/services/hotels-/-restaurants/hotels-open-doors-in-fight-against-virus/articleshow/74974062.cms?utm_source=contentofinterest&utm_medium=text&utm_campaign=cppst

Collins, K. (2020). Bored of life indoors? Take a virtual vacation to the photogenic Faroe Islands. *CNET*, 15 April. https://www.cnet.com/news/bored-of-life-indoors-take-a-virtual-vacation-to-the-photogenic-faroe-islands/

Davies, R. (2020). Disney's Bob Iger stays on to steer company during Covid819 crisis. The Guardian, 13 April. https://www.theguardian.com/film/2020/apr/13/disney-bob-iger-to-stay-as-ceo-during-coronavirus-pandemic

Fox, A. (2020). Hotels are offering up their rooms for healthcare workers thanks to new initiative. *Travel + Leisure*, 24 April. https://www.travelandleisure.com/hotels-resorts/hotels-offer-rooms-ahla-coronavirus

Frenkel, S., Alba, D. & Zhong, R. (2020). Surge of virus misinformation stumps Facebook and Twitter. *New York Times*, 8 March. https://www.nytimes.com/2020/03/08/technology/coronavirus-misinformation-social-media.html

Gill, O. (2020). Branson lashes out after Virgin Australia collapses. *The Telegraph*, 21 April. https://www.telegraph.co.uk/business/2020/04/21/virgin-australia-goes-voluntary-administration-failing-get-bailout/

Gillespie, C. (2020). Restaurants are turning their parking lots into drive-in movie theaters during the pandemic. *Simplemost.com*, 28 April. https://www.simplemost.com/restaurants-turn-parking-lots-into-drive-in-theaters/

Harari, Y.N. (2020). In the battle against coronavirus, humanity lacks leadership. *Time Magazine*, 15 March. https://time.com/5803225/yuval-noah-harari-coronavirus-humanity-leadership/

Harper, J. (2020). Virgin's Richard Branson accused of double standards during coronavirus crisis. *DW*, 22 April. https://www.dw.com/en/virgins-richard-branson-accused-of-double-standards-during-coronavirus-crisis/a-53115031

Hospitality Net (2020a). COVID-19: UNWTO calls on tourism to be part of recovery plans. *Hospitality Net*, 6 March. https://www.hospitalitynet.org/news/4097380.html

Hospitality Net (2020b). Airbnb hosts to help provide housing to 100,000 COVID-19responders. *Hospitality Net*, 26 March. https://www.hospitalitynet.org/news/4097807.html

Hyde, M. (2020). Richard Branson's bailout plea proves there's no one more shameless. *The Guardian*, 21 April. https://www.theguardian.com/commentisfree/2020/apr/21/richard-branson-bailout

Kenny, K. (2020). Coronavirus: With flights grounded indefinitely, how will travel agencies survive COVID-19? *Stuff*, 20 April. https://www.stuff.co.nz/business/121064700/coronavirus-with-flights-grounded-indefinitely-how-will-travel-agencies-survive-covid19

Kwok, L. (2020). Delivery services are getting a boost during the COVID-19recession. *Hospitality Net*, 26 March. https://www.hospitalitynet.org/opinion/4097747.html

Maritime Executive (2020). TUI AG secures government bridge loan to weather COVID-19shutdown. *The Maritime Executive*, 23 April. https://www.maritime-executive.com/article/tui-ag-secures-government-bridge-loan-to-weather-COVID-19-shutdown

Ogden, G. (2020). COVID-19: How the hospitality industry can find a way through. *Hospitality Net*, 17 April. https://www.hospitalitynet.org/opinion/4097842.html

Powley, T., Thomas, D. & Payne, S. (2020). Virgin Atlantic told to resubmit bailout bid by 'unimpressed' UK Treasury. *Financial Times*, 17 April. https://www.ft.com/content/e2636703-acec-49ad-97b8-b944e032db1f

Provident Communications (2019). *Less than half of Canadian businesses are equipped to recover their reputation in a crisis*. 1 May. https://www.newswire.ca/news-releases/less-than-half-of-canadian-businesses-are-equipped-to-recover-their-reputation-in-a-crisis-870690919.html

QSR (2020). Chick-'l-A launches multi-million dollar COVID- 19 relief effort. *QSR Industry News*, 20 April. https://www.qsrmagazine.com/news/chick-fil-launches-multi-million-dollar-COVID-19-relief-effort

Ranosa, R. (2020). Richard Branson: 'I'm working day and night to look after our people'. *HRD Canada*, 23 April. https://www.hcamag.com/ca/news/general/richard-branson-im-working-day-and-night-to-look-after-our-people/220515

Reames, M. (2020). Sportsbooks struggling, Vegas embraces esports betting despite match fixing concerns. *The Washington Post*, 17 April. https://www.washingtonpost.com/video-games/esports/2020/04/17/vegas-esports-betting/

Rizzo, C. (2020). How this Baltimore hotel is helping small businesses, first responders during the coronavirus pandemic. *Travel + Leisure*, 21 April. https://www.travelandleisure.com/hotels-resorts/baltimore-hotel-revival-restaurant-free-space-small-business

Robson, D. (2020). COVID-19. What makes a good leader during a crisis? *BBC Worklife*, 27 March. https://www.bbc.com/worklife/article/20200326-COVID-19-what-makes-a-good-leader-during-a-crisis

Rosen, S. (2020). You can stay at a hotel that uses virus-killing robots to clean its rooms. *The Points Guy*, 26 March. https://thepointsguy.com/news/virus-killing-robot-hotel/

Sumers, B. (2020). What the $2 billion stimulus means for travel businesses. *Skift*, 30 March. https://skift.com/2020/03/30/what-the-2-trillion-u-s-stimulus-package-means-for-travel-businesses/

Toh, M. (2020). Shake Shack returns $10 million emergency loan to the US government. *CNN*, 20 April. https://www.cnn.com/2020/04/20/business/shake-shack-ppp-loan-sba/index.html

Twenty31 (2020). *COVID-19. A strategic approach to address global and regional tourism management challenges*. 3 March. http://www.twenty31.org/thoughtleadership/2020/3/4/COVID-19-a-strategic-approach-to-address-global-and-regional-tourism-management-challenges-march-3-2020

Walton, J. (2020). COVID-19 has severely decreased the number of daily flights. But where are all those grounded planes kept during a pandemic? *BBC News*, 15 April. https://www.bbc.com/worklife/article/20200415-where-are-all-the-unused-planes-right-now

2

3 Crisis communication

Figure 3.1: A typical Micato post on Instagram and Facebook during lockdown

What should I do with my hair these days? It's a question that comes up a lot, so we did a Zoom call this morning with some of our experts on the African plains. Here's one style that can be pulled off at home easily.

Photo: Micato guest Adriane Flinn

Serving close to 5,000 clients a year, Micato Safaris is regarded by many in the luxury travel industry as simply the best in the world at what it does, offering scheduled itineraries and customized trips throughout East Africa, Southern Africa and India. Micato has a well-deserved reputation for giving guests an experience that goes far deeper than any other safari operator, with access to sites rarely open to tourists, while supporting the local economy in myriad ways.

For Micato, relationship marketing has always been a priority, and communicating with customers became even more important during the COVID-19 crisis. As international travel came to a standstill, Micato focused its attention on customer service, reaching out to all customers individually, processing refunds quickly or re-booking vacations. The company also continued with its marketing. "The first thing we did (compared to many of our competitors) was to maintain our marketing budget, although one advantage we have over competitors is that our marketing budget is quite modest. We learned after 9/11 that if you kept your foot on the gas you were stronger coming out of a crisis," said Marty von Neudegg, Executive Director, International for Micato. "In fact, we started paid social media campaigning for the first time during COVID in March – and had fantastic results – I think part of the reason was that competitors had pulled far back or disappeared."

Micato also changed its tone of message, using aspirational language instead of a call to action. "We moved away from phrases like 'book now' and instead said 'start dreaming'. We used powerful images of Africa – 'Moments of Zen' we called them." Using Facebook and Instagram as the social media platforms, the company also posted videos of animals in Africa. "One video about elephants had the highest click through rate in our history, and the open rate was two and half times more than normal. Our pace of enquiries originally fell to about 60% of the same time the previous year, but we are now ahead of last year's pace (raw enquiries). Email traffic is also 20% up from last year. So the tactic is starting to pay off," said von Neudegg in May 2020.

Micato deliberately didn't post anything COVID-related, instead reaching out individually to those who wanted information. "We took a personal approach" said von Neudegg. The company also kept lines of communication open with suppliers and Destination Management Companies (DMCs) in Africa and India. "We have DMCs in Africa and India that solely work with us and do not represent any properties – that gives us an advantage as we can trust them to work in the best interests of our guests – and we had lots of success with most suppliers helping us with refunds or re-bookings. Our partners also gave us up to date information within the borders of the country - South Africa and Kenya have done particularly well with COVID: they had experience dealing with AIDS and Ebola. Those countries had existing protocols in place that helped slow the transmission, including the way they tested all incoming travelers."

In addition to having a reputation for being customer-centric, Micato has always taken corporate social responsibility very seriously. Through the company's One for One Commitment, for every safari they sell, they send a child to school. Micato's nonprofit arm, AmericaShare, also operates a boarding school sponsorship program and provides outreach through its community center in the heart of Nairobi's Mukuru slum. Micato is the only tour company to have won five Condé Nast Traveler World Savers Awards, three for 'Education' and two for 'Doing it All'.

During the COVID-19 crisis, Dennis Pinto, Managing Director for Micato, committed to continue with these significant philanthropic efforts. For example, the company partnered with Huru International in Kenya to supply masks to those in need. Huru, which produces menstrual pads for African women, repurposed some of its production lines and fabric to making face masks for the vulnerable residents of Mukru, one of Nairobi's largest informal settlements with over 300,000 residents. Micato, which has always been a major donor and supporter of Huru, partnered with the nonprofit organization to distribute the masks. By mid-May over 60,000 masks had been distributed to the most vulnerable. At the same time, Huru maintained its production of reusable sanity pads because as they say, "Periods do not stop for Pandemics." "The emphasis on philanthropy through this period was a key driver of Micato's mission to keep going through the crisis," said von Neudegg.

Moving forward, von Neudegg expects that his marketing communications will change as people start to travel again. "We have to decide when we should shift our tone of message to be more sales-oriented. Certainly we will need to reassure our customers that they are in good hands." Operations are also likely to change. "We will review every protocol we can get our hands on and establish best practices for each aspect of our business –airport greetings, lodges and camps, safari vehicle etc. We want the highest standard in each of those. Fortunately, our customers spend most of their vacation with us in big open spaces, and we can offer, for instance, fully private travel, with private helicopter transfers to a private safari lodge, and their only interaction needs to be with their guide and driver. So we are in a good position to come out of this and we are optimistic about the future."

Sources: Personal communication with Marty von Neudegg, May 2020.

3

Introduction

An advertising campaign from Marketing Greece during the COVID-19 pandemic urged international travelers to "#staysafe" during the difficult times, while images of the idyllic Greek Islands allowed viewers to continue to dream and plan an escape to the stunningly beautiful country of Greece. Clever, one might think, sharing the message with the international tourist that better days are surely coming, while urging them to stay safe in the meantime. Yet some suggested that promoting tourism during the pandemic was borderline irresponsible (Spinks, 2020). In fact, quite a few destinations even campaigned against tourism – Visit Wales, for example, urged travelers to stay away. How should the travel industry have responded to this crisis? What was the correct tone of message? And what would persuade travelers to venture out of their homes once the coast was clear? This chapter will explore such issues, in addition to examining internal communications strategies employed by the industry during the pandemic.

The importance of communication during a crisis

Marketing in the midst of a global disaster is always a delicate proposition (Tinubu, 2020). Some organizations choose to go quiet, although previous crises have taught us that marketing is more important than ever at this time (Sapient, 2020). Micato Safaris, profiled in this chapter, appreciated this, having learned from 9/11. Getting the right tone with communications is critical. In North

America, one of the first travel advertisers to pivot away from its usual message strategy was Hotels.com. Rather than focus on vacations or travel rewards, the company quickly launched its 'COVID-19 Social Distancing' ad as soon as the pandemic broke out. It depicted Captain Obvious (the company's mascot) washing his hands and telling viewers 'Just Stay Home'. "We didn't feel the tone of our usual advertising was right for the current environment," said Hotels.com spokesperson Jennifer Dohm. "For the airtime we had remaining, we opted for a message that reinforces the guidance to stay home" (Schaal, 2020). Interestingly, Hotels.com spent more on US national TV in March 2020 than it did during March 2019.

Figure 3.2: Hotels.com - 'COVID-19 Social Distancing' ad

Advertising is a key marketing tool in the tourism and hospitality sector, as providers often require potential customers to base buying decisions upon mental images of product offerings, since they are not able to sample physical alternatives. Often during a crisis or a recession, businesses cut back in their advertising spending – and the COVID-19 pandemic was no exception (Influencer Marketing, 2020). But there have been a number of studies over the years that point to the advantages of maintaining or even increasing advertising budgets during a weaker economy, and there are examples of brands that benefitted by maintaining their ad budgets during economic downturns (Adgate, 2019; Sapient, 2020). There are several reasons to advertise during a slowdown. First, brands can project to consumers the image of corporate stability during challenging times. Second, the cost of advertising drops during a recession. But most important, when you come out of recession your brand is predominant because you have not lost "share of mind" with consumers. During the 1990-1991 recession, for example, Pizza Hut and Taco Bell took advantage of McDonald's decision to drop its advertising and promotion budget. As a result, Pizza Hut increased sales by 61%, Taco Bell sales grew by 40% and McDonald's sales declined by 28% (Adgate, 2019).

During the COVID-19 outbreak, although many brands cut their marketing, there was an increased interest in news, and this sector saw an increase of over 50% in advertising spending. Other areas to see rises in advertising spending were hobbies and interests, technology and computing, and education. Notably, these were all activities that could be consumed at home. Travel advertising saw a rapid decline as can be seen in Figure 3.3, although some organizations in the sector continued to communicate to customers, as will be discussed in the following section.

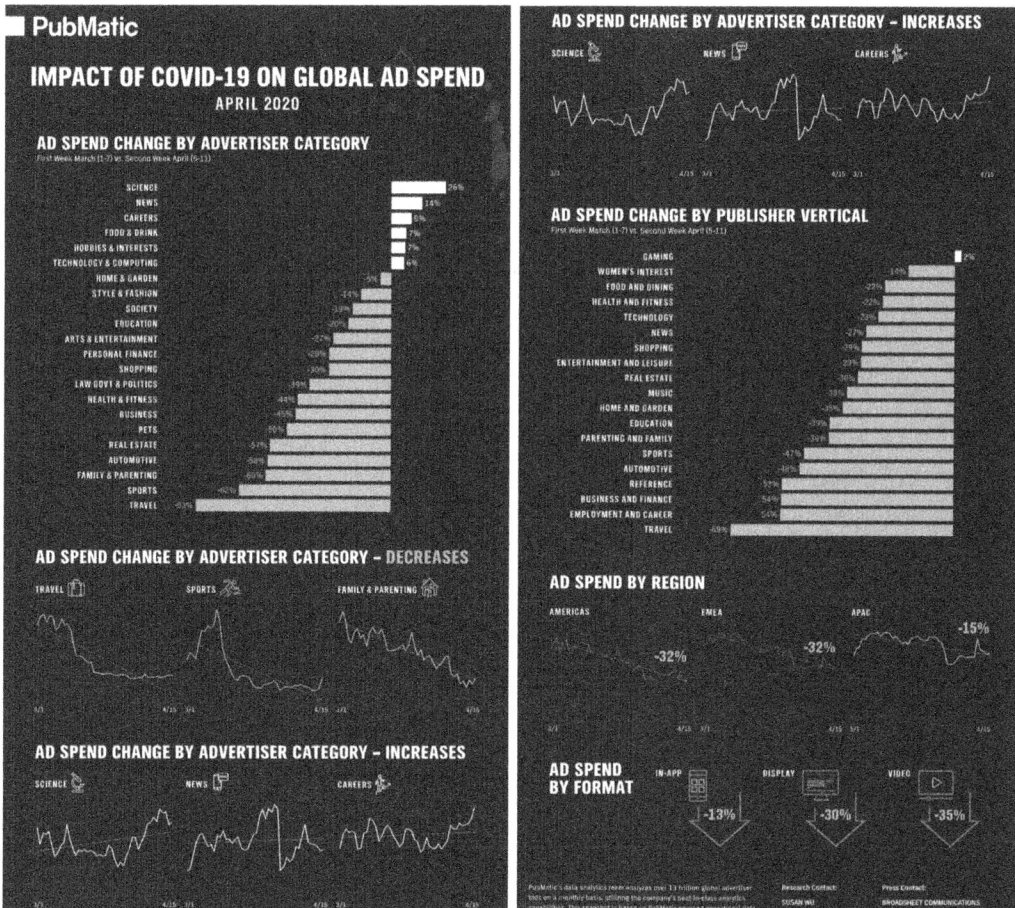

Figure 3.3: Impact of COVID-19 on global ad spend (courtesy of PubMatic, 2020)

Public relations (PR) also plays a key role during any crisis, and we have already seen many examples of PR at work in this book. The goal of PR for an organization is to achieve positive relationships with various audiences (publics) in order to manage effectively the organization's image and reputation. Its publics may be external (customers, news media, the investment community, general public, government bodies) and internal (shareholders, employees). The challenge is to communicate accurately to all stakeholders. During a crisis, well-planned and managed communication could fail if companies privilege some stakeholders over others (Ulmer & Sellnow, 2000). The three most important roles of PR and publicity in tourism and hospitality are maintaining a positive public presence, handling negative publicity, and enhancing the effectiveness of other promotional mix elements (Morrison, 2002). In this third role, PR paves the way for advertising by making customers more receptive to the persuasive messages of ad campaigns. Ultimately, the difference between advertising and public relations is that public relations takes a longer, broader view of the importance of image and reputation as a corporate competitive asset and addresses more target audiences.

Communications strategies used by the travel sector during the crisis

There is a limited number of studies and theoretical models that give guidance to those in the travel industry in terms of communicating during a crisis (Gertner & Kotler, 2004; Hudson, 2016). One can, however, draw on the mainstream marketing and business literature. In the last few decades, research focusing on the media's role during crisis situations and image restoration of companies or places, has led to a variety of models, definitions and concepts, including reputation management, recovery marketing, trust repair, situational crisis communication theory (SCCT) and image restoration. Benoit (1995), for example, lists five communication strategies that can be used in response to a crisis: denial, evasion of responsibility, reducing offensiveness of events, corrective action, and mortification. Coombs (1999) recommends a similar response, although his list extends to seven strategies: attack the accuser, denial, excuse, justification, ingratiation, corrective action and full apology.

Some of the communications from the travel industry in the early stages of the COVID-19 crisis were reactive. Fear and misinformation can spread quickly online, and as mentioned in the previous chapter, this was particularly true of COVID-19. During the pandemic, the Hotel Palatino in Rome came under attack from the media after two guests were found to have coronavirus. While the hotel made every effort to keep guests safe and informed, the story hit mainstream media, resulting in cancelations and bad online reviews. The hotel, therefore, had to make extra efforts to reassure both employees and customers that the hotel was safe. "One of the biggest problems we faced was the media," said Enzo Ciannelli, Hotel Manager. "The spread of fake news or incorrect information fed people's panic and generated a kind of 'mediatic terrorism'. That's why we chose to provide only [basic] and objective information, with the aim of reassuring our customers and employees both offline and online" (Kessler, 2020).

Other communications during the pandemic were more proactive. In April, the World Travel & Tourism Council (WTTC), which represents the travel and tourism sector globally, launched a marketing campaign called TogetherInTravel to galvanize the global travel and tourism community. The marketing campaign comprised three key elements: a highly visual and emotive video; a hashtag – #TogetherInTravel – to stimulate the conversation across social platforms; and a microsite – TogetherInTravel.com – to host the video and user-generated content and stories. Gloria Guevara, president and CEO of WTTC, said: "Dreaming is part of our zest for life and our new campaign encourages thoughts of the brighter days ahead. Our sector builds communities, reduces poverty in the world and improves the social impact of everyday lives. Yet we are uniquely exposed at this time due to COVID-19. Our message is that everyone can still stay inspired with future travel ideas and bookings – and in the meantime be part of a virtual space for sharing, connecting, and collectively inspiring." (Paul, 2020).

Figure 3.4: The World Travel & Tourism Council's TogetherInTravel campaign

For tourism destinations, consultants at Destination Think recommend a transparent communications policy throughout any crisis. During the COVID-19 outbreak, they proposed six elements of crisis planning and online communications

Table 3.1: Crisis communications for destinations during COVID-19 (Destination Think, 2020)

Strategy	Description
1) Identify official sources of information	You can only act appropriately when you have the facts. Determine which news media and government sources your team should depend on for factual updates that can inform your work during the crisis period. Rely on your government's information about the current travel risks, restrictions and recommendations.
2) Place health and safety first	The health and safety of people in your destination always come first. Your social media audiences contain both visitors and residents, so keep local needs in mind, especially if your area reaches a stage where travel is restricted. Support government recommendations about events, gatherings and general movements of the population. Keep messaging consistent within your community.
3) Keep up to date with the news	Make sure your content team stays aware of updates from both the news media and official government sources. Create a plan for responding to events that would trigger a change in messaging, e.g., at what point would you discontinue travel planning and inspirational posts and only post news to keep travelers safe?

4) Be sensitive to the moment	Pause scheduled posts until you've reviewed them in light of the current situation. Identify messages that your audience might interpret as insensitive, and avoid posts that may inadvertently make people think of the virus instead of the destination. Change copy or images if you are concerned that they might draw criticism or negative comments.
5) Be transparent and informative	If major attractions close or large events cancel, be transparent. You may be able to offer alternative suggestions. Your team needs to be able to answer questions and respond to comments about the status of popular tourism draws. Be direct, but sympathetic in tone. Point people to official sources. Acknowledging a traveler's hardship or lost opportunity while providing facts can defuse a tense situation.
6) Monitor and moderate	Be present on your social media channels. Depending on your community's mood and response, your team may need to devote extra time here. As always, take note of common questions and correct misinformation by responding to comments with facts. By responding to individual concerns and avoiding canned messages, travelers will feel heard and valued.

As they recommend, organizations need to be present on social media channels, and in recent years, social media has become a critical component of crisis communication. Cheng (2008) suggests that monitoring, analyzing and understanding stakeholders' needs and desires in crises might be the first step before making any decisions about crisis responses. Social media can play an important monitoring role at this stage (Eriksson, 2018). By tracing online data such as 'likes', 'links' and 'organic views', and already-censored growing negative emotions or information needs shared from stakeholders, crisis managers can immediately evaluate response to a crisis and adopt the right strategy to manage the situation. Interestingly, one report showed that social media engagement on Twitter during the crisis saw more of an increase than on Instagram or Facebook, suggesting that people were turning to Twitter for critical information and news, and then retweeting and discussing these topics (Influencer Marketing, 2020).

Social media can also be a channel for keeping consumers informed during a crisis. Destinations like Salzburg, Austria, for example, announced closures in their resort through social media platforms by placing a clear priority on visitor health and by expressing sympathy to those whose plans had been disrupted. A company's website is also another trusted source of information during a crisis. Some organizations updated their homepages with the most relevant news and information on COVID-19 using a temporary page banner or a news section on their site. Auckland Tourism, Events and Economic Development had both, posting a temporary page banner on its homepage called 'Tamaki Makaurau Auckland – COVID-19 updates' and a link on its homepage to 'Latest News'. More details about this can be found in the end of chapter case study. Other destinations created temporary banners for their social media sites. For example, the Facebook cover picture for the Town of Canmore in Canada (the destination where the author wrote this book!) was COVID-related, using the somewhat hackneyed phrase "We are in this together" (see Figure 3.5).

Figure 3.5: Canmore's temporary Facebook cover image during the crisis

Other organizations, because it was an ongoing crisis, created an additional dedicated page that was search engine optimized, and consistently updated this with the latest news. This helped to counter any negative news media that might have shown up when potential travelers were searching a specific company on Google. Like social media, websites require continual updating, but they are a useful resource and can increase consumer confidence.

A number of tourism destinations used social media to keep their brands top of mind during the crisis as we saw in the opening case study. One of the first DMOs to do this was Marketing Greece (mentioned previously). The campaign used different photographs accompanied by the caption 'When the time is right, we'll be there for you. Till then #staysafe', using the unique crystal-clear light of Greece to bring hope to the quarantine-weary public. The overall aim was to share the message with international tourists that better days were surely coming, while urging them to stay safe in the meantime. A spokesperson from Marketing Greece said "In our times, humanity is called upon to rise to a shocking challenge, with messages of hope and optimism being more imperative than ever. Greek tourism, fully identifying with the sense of freedom and escape from the everyday grind, sends its own message for the next day."

Other destinations followed a similar strategy, teasing the beauty of their respective locations, but encouraging visitors to stay home for now. Alaska's website featured a video showcasing glaciers and mountains and promising that 'Alaska will wait, for you'; San Diego said 'We'll keep San Diego warm for you'; Visit Scotland said 'Absence Makes the Heart Grow Fonder'; Kenya created a YouTube and Instagram campaign that used intriguing images of the country to remind travelers that all these things would be waiting for them when the pandemic passes; Turkey's #TogetherToday print campaign urged everyone to stay home 'so we can explore again tomorrow'; and Portugal's emotional Instagram campaign stressed the need for social isolation at this time, while portraying breathtaking images from across the country.

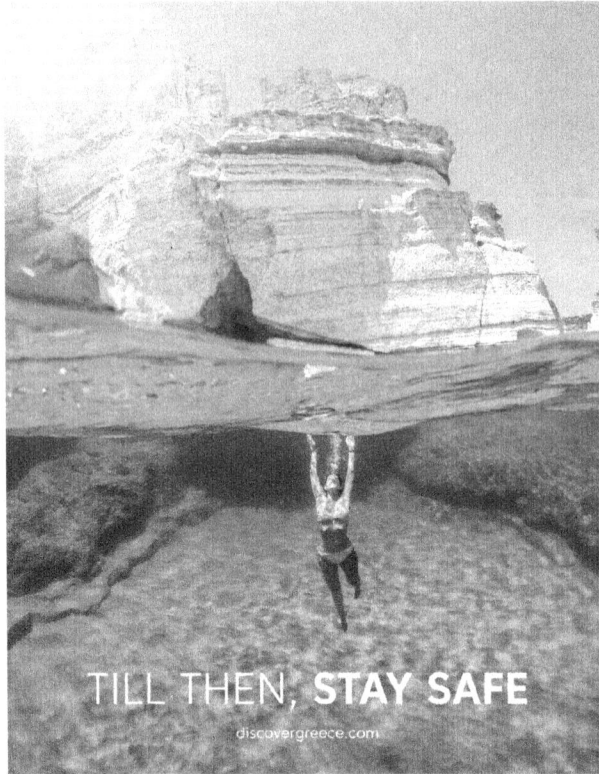

Figure 3.6: Marketing Greece's 'till then #staysafe' campaign (courtesy of Marketing Greece)

Figure 3.7: The Las Vegas #OnlyYou campaign (courtesy of LVCVA)

Visit Las Vegas, with its #OnlyYou campaign (Figure 3.7) showcased an empty Las Vegas Strip, reminding people that the city would be there when they started to travel again. "After coming off a successful launch of 'What Happens Here, Only Happens Here' campaign which first aired on January 26, 2020, we quickly had to pivot messaging and create a spot to fill our national TV buy in the U.S. and Canada," said Colleen Yoshida, Senior Director of Advertising at the Las Vegas Convention and Visitors Authority (LVCVA). The #OnlyYou spot was created over a weekend and, in addition to running on broadcast TV, the ad ran on Connected TV, pre-roll and YouTube. In addition, LVCVA received donated space from media partners because of the relevant message. "We also translated the subtitles and shared it with international partners in six different languages," said Yoshida.

Other campaigns suggested that the world was getting a well-needed rest and would be even better when we could travel again. Tourism Australia, for example, released a video called 'With Love From Aus', emphasizing the fact that nature was getting some rest and relaxation while tourists were in isolation. The video combined webcam shots of people staying connected, with images of Australian landscapes and native animals like the quokka, koala and 'roos running amok'. Tourism Australia Managing Director Phillipa Harrison said that through these times of global uncertainty, everyone could do with a bit of positivity and something to look forward to: "With Love From Aus is a heartfelt message to all Australians and the rest of the world that our beautiful country and its people will be ready to welcome visitors with open arms when the time is right." Auckland in neighboring New Zealand also hit the headlines with an inspirational video, and this is profiled in the end of chapter case study.

On a similar vein, Abu Dhabi's Department of Culture and Tourism posted a video that began with 'The world is quieter now', before launching into a series of images showing the beauty of the capital. It ended by telling viewers that things would revert to normal soon, and that Abu Dhabi would be waiting for visitors with open arms when that happens. The video was produced both for potential visitors and for residents. Saint Lucia followed a different strategy by streaming live footage on Instagram Live. Saint Lucia's Tourism Authority showcased seven 'virtual' minutes on the Caribbean island. Aired twice a week, the mini-series started with a seven-minute yoga session under the shadow of the volcanic Piton spires. Island vibes came courtesy of Caribbean cookery classes, beach meditation and SocaFit classes. For podcast fans, there was also a new five-part armchair guide to the island, ideal for anyone dreaming of Saint Lucia during self-isolation.

Zermatt in Switzerland took a subtler but still effective approach to keep their destination top of mind. Every night during the crisis, from March 24 to April 26, they projected different symbols on to the iconic Matterhorn mountain. The village wanted to show solidarity with all those who were suffering around the world and send gratitude towards everyone helping to overcome the crisis. One night they projected the British flag (Figure 3.8), and on a Facebook post said: "Dear friends from the United Kingdom, Zermatt is deeply connected to you

Figure 3.8: Zermatt in Switzerland projecting the British flag on to the Matterhorn (courtesy of Zermatt Tourismus: Light Art by Gerry Hofstetter, Photo Gabriel Perren)

through its history. British mountaineers first brought tourism to Zermatt, and to this day, Zermatt can welcome many visitors from the UK. Unfortunately, travelling is not possible right now. We send you a sign of hope from the Matterhorn instead." The campaign generated a considerable amount of media worldwide, reaching over 700 million people, with postings on social media receiving up to 1.7 million likes. Opinion leaders around the globe shared the images of the illuminated Matterhorn in their social media channels – including Narendra Modi (Prime Minister of India), Sebastian Kurz (Austrian Chancellor), Alain Berset (Swiss Federal Councillor), Lee Hsien Loong (Prime Minister of Singapore), Stephen Colbert (US satirist) and the Indian actors Ranveer Singh, Katrina Kaif and Anushaka Sharma.

As countries began talking about easing lockdown laws, some destinations reached out to travelers with more direct calls to action. For example, the Italian island of Sicily made the headlines late in April 2020 when its regional government said that once borders were open, it would cover half of flight costs and a third of hotel expenses for travelers wishing to visit, as well as free tickets to many of its museums and archaeological sites. The island had lost over $1 billion in tourism-related revenue after having to close, and while the proposed plan would cost the island over $50 million, it hoped to make this back as lockdown laws eased. About the same time, Secretary of State for Tourism in Portugal, Rita Marques said the country would give vouchers to tourists forced to cancel their holiday plans because of the pandemic, allowing them to reschedule trips through to the end of 2021. "We are being absolute pioneers in the European context. Our priority is to safeguard consumer rights and the interests of economic operators, according to the principle of 'don't cancel, postpone'," said Marques (Goncalves, 2020).

Communications post-crisis

As the lockdown period came to an end, destination marketers began to think about strategies to bring back the tourists. Niagara Falls Tourism, for example, was working hard in April 2020 on recovery campaigns "to make sure we have a clear, compelling message when the time comes," said Janice Thomson, the organization's President and CEO. She added that the messaging may focus on outdoor experiences to reduce the potential for crowding. "It was like a switch was turned off when this crisis started, and I think it's going to be like a switch being turned back on once we have clearance from public health authorities" (Bisby, 2020).

For many tourism destinations, for example Italy and Spain, the challenge after the pandemic will be repairing a tarnished image. Avraham & Ketter (2008) have proposed a conceptual framework called the 'multi-step model for altering place image', which offers three types of strategies that destinations can follow to promote their destinations during and after tourism crises: source strategies, message strategies, and audience strategies. Several studies have successfully used this model, or part of it, to analyze the marketing efforts of destinations around the world following a crisis (Avraham, 2015; Avraham & Ketter, 2016; Walters & Mair, 2012).

Source strategies concentrate mainly on the marketers' efforts to influence or replace the source that is perceived as responsible for the destination's negative image, usually the media. Two examples of this are organizing familiarization tours for journalists (Mair et al., 2016) and launching a social media campaign to bypass the news media (Ketter & Avraham, 2012). **Message strategies** focus on tackling the negative messages reported about the destination by sending opposite messages, reducing the scale of the crisis or expanding the destination image beyond the stereotype attached to it. Kyoto in Japan followed this strategy at the beginning of the pandemic by launching an 'empty tourism' campaign to lure travelers back to the city (Jozuka, 2020). Finally, **audience strategies** are concerned with the audience's values and perceptions. Here, marketers try to highlight some of the values held in common between their destination and the target audience.

Towards the end of the lockdown, VisitScotland followed an audience strategy by launching a campaign called #AWindowOnScotland. Realizing that domestic tourism would be the first to recover, the idea was to create an authentic picture of the country during the lockdown by opening a window on to people's experiences while they stayed at home. The campaign encouraged people to share views from their homes, and popular posts included images of the coastline at Oban, a video of lawn-mowing at lighthouse cottages on Orkney, the sunset from Gourock, and a video from a lodge on the banks of Loch Ness. The #AWindowOnScotland campaign was part of a national tourism strategy to help recover the economy when the coronavirus lockdown was lifted. Staycations and day visits would be promoted when restrictions began to ease, as part of a national action plan from the Scottish Tourism Emergency Response Group. Malcolm Roughead, Chief Executive of VisitScotland said: "We're thrilled at the reaction to #AWindowOnScotland from residents, and heartened businesses are engaging with the campaign and using the opportunity to share their windows on Scotland with the world during this challenging time. The staycation market will be a key driver in the country's economic recovery and the many wonderful views people have been sharing while they stay at home will surely be an incentive for many to travel when the time comes" (Herald Scotland, 2020).

Cause-related marketing

As mentioned above, one of the reasons to keep lines of communication open during a crisis is to project to consumers the image of corporate stability during challenging times. But organizations will also want to portray an image of corporate responsibility, and the COVID-19 crisis presented them with the opportunity to engage in cause-related marketing (CRM). CRM is a rapidly expanding trend in marketing communications, and is basically a marketing program that strives to achieve two objectives – improve corporate performance and help worthy causes – by linking fundraising for the benefit of a cause to the purchase of the firm's products and/or services (Barnes & Fitzgibbons, 1991). Companies use CRM to contribute to the well-being of society and to associate themselves with a respected cause that will reflect positively on their corporate image. Companies, and their brands, can benefit from strategic alignments with causes or with not-for-profit organizations. It is hoped the emotional attributes associated with cause-linked brands differentiate them from their rivals (sometimes referred to as 'cause branding').

CRM efforts can be categorized into autonomously branded, co-branded, house branded, and industry branded approaches (Hudson, 2008), the distinctive features of each being summarized in Table 3.2. **Autonomously branded** philanthropic collaborations are characterized by an arm's length relationship between the corporate sponsor's brand and the brand of the charity/cause that it supports. These are the quickest and easiest kind of relationships to arrange and are the dominant form of philanthropic activity. During the COVID-19 crisis, for example, Marriott Bonvoy members could donate points to relief organizations, including the American Red Cross, International Federation of Red Cross and Red Crescent Societies, UNICEF, and World Central Kitchen. Additionally, through its Community Caregiver Program, Marriot was offering special rates for first responders and healthcare professionals who wanted to book rooms near hospitals.

Table 3.2: Branding approaches to CRM

Features	Autonomously branded	Co-branded	House branded	Industry branded
Charity reputation	Established/ independent of company	Tied to company and charity	Company dependent	Established/ independent of industry
Company's involvement in charity administration	None	Partial to jointly administered	Company controlled	None
Company control over charity use of funds raised by CRM Program	Limited	Some influence	Complete	Limited
Strategic opportunity	Leverage external brand	Leverage brand congruence	Support existing company/ product brand	Leverage industry brand
CRM promotional objective	Demonstrate firm-charity congruence where not obvious	Demonstrate firm-charity brand congruence	Promote firm's commitment	Promote industry's commitment

In **co-branded** collaborations, the company and the charity join together for a new charitable cause. An example of this type of branding approach is the partnership between Micato Safaris and Huru International in Kenya referred to in the opening case study. During the crisis, Huru, which produce menstrual pads for women in Africa, repurposed some of its production lines and fabric to making face masks for the vulnerable residents of Mukru, one of Nairobi's largest informal settlements with over 300,000 residents. Micato, partnered with the nonprofit organization to distribute over 50,000 masks by May 2020.

With the **house branded** approach, the firm takes ownership of a cause and develops an entirely new organization to deliver benefits associated with the cause (Hoeffler & Keller, 2002). Similar to private label products, house branded charities are by definition differentiated from other charities, and in the increasingly crowded marketplace for philanthropic program partners, the firm has unfettered access to its own charity. Vail Resorts in Colorado has its own house branded charity called EpicPromise Employee Foundation which helps the company's employees respond to unpredictable setbacks, including medical events. During the COVID-19 crisis, CEO Rob Katz donated an additional $1 million to the fund to help meet the increased need for assistance. "I cannot recall another moment in my lifetime that has caused so much disruption to our lives – to our work, to our health and to our communities," said Katz. "Throughout this incredibly challenging time, two of our absolute priorities have been, and will continue to be, the health and well-being of our employees and mountain communities."

AHLA'S
HOSPITALITY
FOR HOPE
INITIATIVE

Hotels Supporting Healthcare:
COVID Toolkit

*Industry resources to support the health care community,
first responders, displaced employees, and local
communities during the crisis*

AHLA
AMERICAN HOTEL & LODGING ASSOCIATION

Figure 3.9: AHLA's COVID-19 toolkit

Finally, **industry branded** initiatives are those that involve contributions to a cause from the industry as a whole, rather than separate corporations. An example of this type of CRM is the 'Hospitality for Hope' initiative mentioned in Chapter 2 that was launched by the American Hotel and Lodging Association (AHLA). The program identified hotel properties that had volunteered to provide temporary housing for emergency and healthcare workers. Participating hotels could potentially be used as emergency hospitals or housing for those who required quarantine. Over 16,000 hotels had signed up by mid-April 2020. AHLA provided a 'COVID-19 Toolkit' that offered industry advice on how to support the health care community and first responders, displaced employees, and local communities during the crisis.

3

Internal communication

Internal communication plays a key role during a crisis. Employees should be up to date on the latest news, any crisis plan procedures, and ready to deal with 'what if' scenarios. They should also have the opportunity to voice their own concerns and ask follow-up questions (Huang, 2020). Previous research has shown that effective communication during a crisis needs to be consistent, timely and active (Huang & Su, 2009; Mazzei & Ravazzani, 2011; Strong, River & Taylor, 2001; Coombs, 2018). However, owing to the uncertainty related to a crisis event, multiple interpretations can emerge, creating a situation of high communication ambiguity. There can, therefore, often be a missing link during a crisis between what communicators intend to communicate and what employees actually perceive (Mazzei & Ravazzani, 2011).

To cope better with such complexities and reduce misalignment during a crisis, it has been suggested that internal communication practitioners should first develop deep trust relationships with employees before a crisis occurs (O'Hair et al., 1995). Then, actions and factual communication that express concern for employees should accompany any formal declaration to give credibility and consistency to messages (Benoit, 1997). Communication with employees can be one-to-one, via company meetings, training sessions, webinars, newsletters, emails, annual reports, or video recordings. During the COVID-19 crisis, Virgin Founder Richard Branson decided to publish a lengthy open letter to his employees to respond to all the criticism he was receiving from governments and media (see the case study in Chapter 2). Addressed to the 'Virgin family' he said "much has been said about me and our brand" and that it was "important for you all to know the actual facts". Branson said he was "working day and night to look after our people and protect as many jobs as possible", despite working in one of the "hardest hit sectors, including aviation, leisure, hotels and cruises". "We're doing all we can to keep those businesses afloat and I am so thankful to all of you who have continued to work so hard in these difficult times," he said.

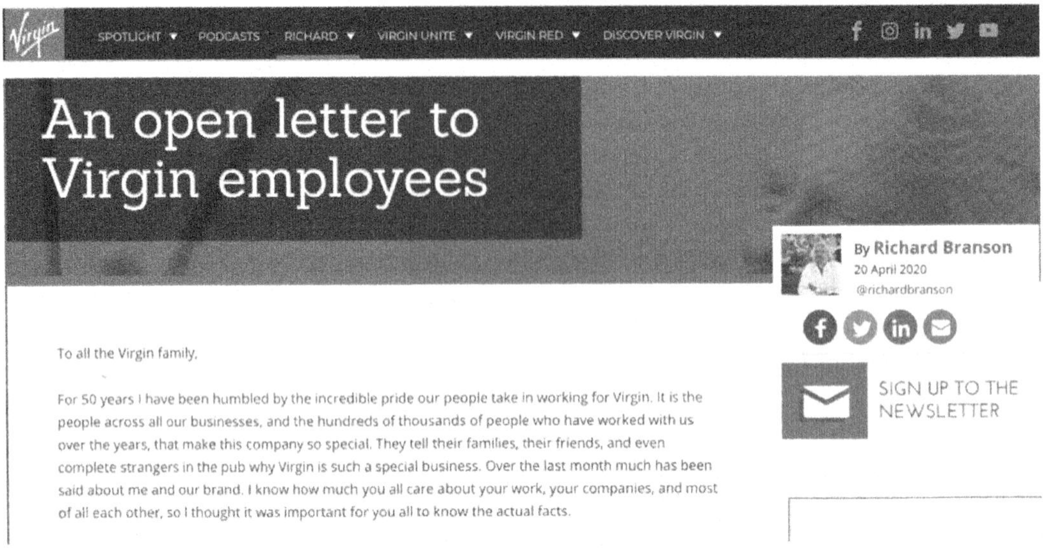
SPOTLIGHT ▾ PODCASTS RICHARD ▾ VIRGIN UNITE ▾ VIRGIN RED ▾ DISCOVER VIRGIN ▾

An open letter to Virgin employees

By Richard Branson
20 April 2020
@richardbranson

SIGN UP TO THE NEWSLETTER

To all the Virgin family,

For 50 years I have been humbled by the incredible pride our people take in working for Virgin. It is the people across all our businesses, and the hundreds of thousands of people who have worked with us over the years, that make this company so special. They tell their families, their friends, and even complete strangers in the pub why Virgin is such a special business. Over the last month much has been said about me and our brand. I know how much you all care about your work, your companies, and most of all each other, so I thought it was important for you all to know the actual facts.

Figure 3.10: Virgin open letter to employees from Richard Branson

Marriott CEO Arne Sorenson took to social media to address his employees during the crisis. In a video released on Twitter, Sorenson committed to forgoing a salary for the remainder of the year and cutting those of his executive team by 50%. Sorenson said that the financial impact of the crisis was more severe than 9/11 and the 2008 financial crisis combined. "The worst quarter we had in those earlier crises saw a roughly 25% decline in hotel revenues on average across the globe. In this case, which began in Greater China in January, we quickly saw a 90% decline in business in China." Sorenson added that, in most markets, business was running 75% below normal levels. He told employees that the company had suspended all non-essential travel and spending, new hires outside of "mission-critical positions", hotel initiatives for 2020, and brand marketing.

Thomas Willms, CEO of Deutsche Hospitality, said that during the crisis his company was offering training webinars to employees while they were in lockdown. Deutsche Hospitality operates five hotel brands on three continents, and the group's 'Staff Training Staff' project won an eLearning Award in 2020. Employees prepare brief training videos on selected topics relating to their daily work and they are then made available to their co-workers. Videos can be accessed via mobile devices at any time and in any location. "Morale is key," Willms said in an interview during the crisis. "And we are committed to maintaining jobs. We do video messaging, we have online training seminars, and we communicate reopening schedules to keep their morale up." Willms was speaking on a podcast put on by QUO, a hospitality-*branding* agency based on Bangkok, which produced a series of podcasts on *The Future of Travel* during the crisis, interviewing key leaders from the travel industry – https://www.quo-global.com/podcasts/. Another smart example of keeping a brand top of mind!

To keep up morale in Las Vegas, the city thanked employees on the frontlines of the hospitality industry during National Travel and Tourism Week (NTTW) May 3-9, 2020. This is an annual celebration recognizing the contributions of employees in the US travel industry. "At the heart of the travel industry are the dedicated hospitality workers, and we appreciate all they do to make Las Vegas the world's best destination," said Steve Hill, LVCVA president and CEO (LVCVA, 2020). Las Vegas honored NTTW with the release of a video on Sunday May 3, featuring poignant messages from local hospitality workers. During a 'Red Takeover', various resorts and attractions from downtown to the Las Vegas Strip – including the iconic 'Welcome to Fabulous Las Vegas' sign – were lit red, the official color of NTTW, in an evening show of solidarity.

3

Figure 3.11: The iconic 'Welcome to Fabulous Las Vegas' sign was lit red during National Travel and Tourism Week (courtesy of LVCVA)

Finally, as the industry tentatively started back up again after lockdown, changes in the service delivery process had to be communicated to stakeholders and employees in a timely fashion. In Portugal, Turismo de Portugal created a new 'Clean & Safe' certification for the country's hotels, tourism enterprises, entertainment companies and travel agencies. Tourism organizations could apply for this stamp of approval to show that they were compliant with hygiene and cleaning requirements for the prevention and control of COVID-19 and other possible infections. The idea was to reinforce visitors' confidence in the safety of the destination. Portugal's Directorate-General for Health (DGS, the national health authority) created online training programs for employees in the industry, including providing information to guests and offering protective equipment like masks. "We realized from the beginning of this crisis that we needed to support three communities: travel companies, tourists and citizens," said Luís Araújo, President of Turismo de Portugal, when asked about the reasons for Clean & Safe. He said the lockdown period "gave us time to build trust and prepare for the new normal" (Abel, 2020).

Case study: Keeping lines of communication open in New Zealand

Figure 3.12: Auckland's 'Papatūānuku is breathing' campaign (courtesy of ATEED)

One organization that was very proactive in communicating to all of its stakeholders during the COVID-19 crisis was Auckland Tourism, Events and Economic Development (ATEED). ATEED is the region's economic development agency, and during the lockdown period it hit the headlines after producing an inspirational video. The video, *Papatūānuku is breathing*, was narrated by 11-year-old Manawanui Makiapoto Mills and began by panning across Kariotahi Beach on Auckland's west coast. "Stop. Listen. Papatūānuku, the earth mother, is breathing. Tāmaki Makaurau, Auckland. Still," Mills says. The video ends with: "And when the time is right, we welcome you. But for now, listen. Papatūānuku is breathing." Aware that other destinations around the world had created showcase videos amid the COVID-19 crisis, ATEED wanted to come up with something that was uniquely Auckland.

Papatūānuku is breathing went viral around the world. Within three weeks, it had been translated into seven languages by inspired viewers and had been viewed more than a million times on Visit Auckland's social media channels alone. "We are proud of the positive feedback that we have received about our video. That will provide us with inspiration as we move more purposefully alongside our tourism industry towards recovery," said Steve Armitage, ATEED's General Manager. "We are grateful to all those who collaborated with us on this very special project and gifted us the use of their material." Mayor Phil Goff said the video was an inspiring portrayal of Auckland's beautiful natural environment. "As we all play our part in the fight against COVID-19 by staying home and staying local, *Papatūānuku is breathing* is a heartening glimpse of what we have to look forward to when the lockdown is lifted and a showcase of what our city has to offer," he said.

To leverage the success of the video, the team used user generated content (UGC) and excerpts from the *Papatūānuku is breathing* script to carry the sentiment through their social media posts. But producing this video was only one part of ATEED's communications strategy during the COVID-19 outbreak. For up to date information on COVID-19 related subjects, ATEED posted a temporary page banner on its homepage called 'Tamaki Makaurau Auckland – COVID-19 updates' and provided up to date news articles on its 'Latest News' link. The organization was also very active on social media, posting regularly on Facebook, YouTube, LinkedIn and Instagram. Weekly email updates to thousands of Auckland businesses alerted them to practical support available from regional and national sources, and highlighting updates to the resources page.

During the crisis, ATEED surveyed businesses to gauge how the initial response to COVID-19 was affecting them and how they were reacting. Financial and cash flow management advice and business strategy and planning were the top types of support Auckland businesses wanted as the response to COVID-19 ramped up. Increased digital marketing and social media capability and help targeting new markets, with a focus on local and domestic customers, were also high on the list. ATEED used insights gained from the survey to shape the business support it provided, to ensure it was delivering information and resources local businesses wanted. A new resources webpage, informed by the survey response, collated useful tools and information to help businesses of all sizes through the crisis. This page was being updated regularly.

ATEED also reached out to small and medium-sized businesses offering a free online tool to help them boost their digital know-how, including guidance on remote working, e-commerce, and increasing their online presence to make it easier for customers to find them. The tool was developed in partnership with social enterprise Digital Journey. "Our small to medium enterprises are a key part of the region's economy and employment," said Mayor Phil Goff. "This online tool will provide practical assistance to businesses looking to adapt to the new realities of working under the COVID-19 alert system, which in many cases will require rapid adoption of digital and online technologies."

Finally, ATEED partnered with the Employers' & Manufacturers' Association (EMA), and the Regional Business Partner Network to launch a free web series to help all businesses get through the COVID-19 pandemic and beyond. The combination of videos and live webinars covered topics such as mental well-being for employers, retention and redeployment, employee leave and payroll. Each week, three 20-minute sessions were added to the series on the topics that businesses were most asking about. As an outcome of all these initiatives, community engagement was very positive. In its first week, 427 businesses used the digital assessment tool, completing assessments and getting customized action plans. This exceeded its aim of 400 businesses over three months, which was the project target set pre-COVID-19.

Sources: Doyle (2020); Personal communication with Chris Gregory, May 2020

■   References

Abel, A. (2020). Hotels and travel in Portugal are getting safer from the coronavirus. *Forbes*, 6 May. https://www.forbes.com/sites/annabel/2020/05/06/heres-how-hotels-and-travel-in-portugal-are-getting-safer-from-covid-19/#2e3dad092bb2

Adgate, B. (2019). When a recession comes, don't stop advertising. *Forbes*, 5 September. https://www.forbes.com/sites/bradadgate/2019/09/05/when-a-recession-comes-dont-stop-advertising/#3ce95afc4608

Avraham, E. (2015). Destination image repair during crisis: Attracting tourism during the Arab Spring uprisings. *Tourism Management*, **47**, 224–232.

Avraham, E. & Ketter, E. (2008). *Media Strategies for Marketing Places in Crisis. Improving the image of cities, countries and tourist destinations.* Oxford, UK: Butterworth-Heinemann.

Avraham, E. & Ketter, E. (2016). *Marketing Tourism for Developing Countries: Battling stereotypes and crises in Asia, Africa and the Middle East.* London: Palgrave-McMillan.

Barnes, N.G. & Fitzgibbons, D. (1991). Is cause related marketing in your future? *Business Forum*, **16**(4), 20.

Benoit, W.L. (1995). *Accounts, Excuses and Apologies: A theory of image restoration strategies.* Albany, NY: State University of New York Press.

Benoit, W.L. (1997). Image repair discourse and crisis communication. *Public Relations Review*, 23(2), 177-86.

Bisby, A. (2020). How the travel industry is keeping tourists engaged while waiting to welcome them again. *The Globe & Mail*, 31 March. www.theglobeandmail.com/life/travel/article-how-the-travel-industry-is-keeping-tourists-engaged-while-waiting-to/

Cheng, Y. (2018). How social media is changing crisis communication strategies: Evidence from the updated literature. *Journal of Contingencies and Crisis Management*, **26**, 58-68.

Coombs, W.T. (1999). Information and compassion in crisis responses: A test of their effects. *Journal of Public Relations Research*, **11**(2), 125-42.

Coombs, W.T. (2018). *Ongoing Crisis Communication: Planning, managing and responding.* California: Sage.

DestinationThink (2020). COVID-19 pandemic needs rational leadership from tourism destinations. https://destinationthink.com/blog/COVID-19-pandemic-rational-leadership-tourism-destinations/

Doyle, A. (2020). Video brings Auckland to the world during lockdown. *Recommend Magazine*, 20 April. https://www.recommend.com/amazingdaysahead/video-brings-auckland-world-lockdown/

Eriksson, M. (2018). Lessons for crisis communication on social media: A systematic review of what research tells the practice. *International Journal of Strategic Communication*, **12**(5), 526-551.

Gertner, D. & Kotler, P. (2004). How can a place correct a negative image? *Place, Branding and Public Diplomacy*, **7**(2), 50-57.

Goncalves, C. (2020). "Don't cancel, postpone," Portugal urges tourists in voucher scheme. *Reuters*, 23 April. www.reuters.com/article/us-health-coronavirus-portugal-touri...el-postpone-portugal-urges-tourists-in-voucher-scheme-idUSKCN2252OT

Herald Scotland (2020). Scottish tourism campaign a massive hit. *Herald Scotland Online*, 8 May. https://www.heraldscotland.com/news/18437396.scottish-tourism-campaign-massive-hit/

Hoeffler, S. & Keller, K.L. (2002). Building brand equity through corporate societal marketing. *Journal of Public Policy & Marketing*, **21**(1), 84.

Huang, N. (2020). Crisis planning for hotels: Best practices in communication, revenue management, and marketing. *Hospitality Net*, 4 March. https://www.hospitalitynet.org/opinion/4097317.html

Huang, Y.-H. & Su, S.-H. (2009). Determinants of consistent, timely and active responses in corporate crises. *Public Relations Review*, **35**(1), 7-17.

Hudson, S. (2008). *Tourism and Hospitality Marketing: A Global Perspective*. London: Sage.

Hudson, S. (2016). Let the journey begin (again). The branding of Myanmar. *Journal of Destination Marketing & Management*, **5**(4), 305-313.

Influencer Marketing (2020). Coronavirus (COVID-19) Marketing & Ad Spend Impact: Report + Statistics. https://influencermarketinghub.com/coronavirus-marketing-ad-spend-report/

Jozuka, E. (2020). Kyoto launches an 'empty tourism' campaign amid coronavirus outbreak. *CNNI*, 18 February. https://www.msn.com/en-us/travel/news/kyoto-launches-an-empty-tourism-campaign-amid-coronavirus-outbreak/ar-BB106A2h

Kessler, M. (2020). 6 ways to manage reputation during coronavirus & other crises. *Hospitality Net*, 5 March. https://www.hospitalitynet.org/opinion/4097310.html

Ketter, E. & Avraham, E. (2012). The social revolution of tourism marketing: The growing power of users in social media tourism campaigns. *Place Branding and Public Diplomacy*, **8**(4), 285–294.

LVCVA (2020). Las Vegas thanks hospitality workers and honors the spirit of travel. LVCVA Press Release, 1 May. https://press.lvcva.com/news-releases/all/las-vegas-thanks-hospitality-workers-and-honors-the-spirit-of-travel/s/d4b43220-27b2-4b58-b894-06447d2c7c3d

Mair, J., Ritchie B.W. & Walters G. (2016). Towards a research agenda for post-disaster and post-crisis recovery strategies for tourist destinations: A narrative review. *Current Issues in Tourism*, **19**(1), 1–26.

Mazzei, A. & Ravazzani, S. (2011). Manager-employee communication during a crisis: The missing link. *Corporate Communications: An International Journal*, **16**(3), 243-254.

Morrison, A.M. (2002) *Hospitality and Travel Marketing*, 3rd edn. Albany, NY: Delmar Thomson Learning.

O'Hair, D., Friedrich, G.W., Wiemann, J.M. & Wiemann, M.O. (1995). *Competent Communication*. New York, NY: St Martin's Press.

Paul, M. (2020). WTTC unveils 'Together in Travel' campaign. *Travel Daily*, 21 April. https://www.traveldailymedia.com/wttc-unveils-together-in-travel-campaign/

PubMatic (2020). The impact of COVID-19 on global ad spend. *PubMatic*, 15 April. https://pubmatic.com/COVID-19/

Sapient, P. (2020). Preparing for a post COVID-19 world: 4 ways travel brands can learn from past events. *Skift*, 9 April. https://skift.com/2020/04/09/preparing-post-COVID-19-world-travel-brands-learn-past-events/

Schaal, D. (2020). How travel brands are approaching TV advertising now. *Skift,* 27 March. https://skift.com/2020/03/27/how-travel-brands-are-approaching-tv-advertising-now/

Spinks, R. (2020). Promoting tourism in the time of Coronavirus is a no-win. *Skift*. 5 March. https://skift.com/2020/03/05/promoting- tourism-in-the-time-of-coronavirus-is-a-no-win/

Strong, K.C., Ringer, R.C. & Taylor, S.A. (2001). The rules of stakeholder satisfaction (timeliness, honesty, empathy). *Journal of Business Ethics*, **32**(3), 219-30.

Tinubu, S. (2020). 5 things advertisers should consider amidst the COVID-19 pandemic. *Entrepreneur*, 29 April. https://www.entrepreneur.com/article/348526

Ulmer, R.R. & Sellnow, T.L. (2000). Consistent questions of ambiguity in organizational crisis: Jack in the Box as a case study. *Journal of Business Ethics*, **25**, 143-55.

Walters, G. & Mair, J. (2012). The effectiveness of post-disaster recovery marketing messages – the case of the 2009 Australian Bushfires. *Journal of Travel & Tourism Marketing*, **29**(1), 87–103.

4 How travelers were affected by the COVID-19 crisis

Figure 4.1: Ed and Barbara Sobey

For Ed and Barbara Sobey, Pagodas and Pearls was their winter escape from the soggy Pacific Northwest. The all-inclusive luxury cruise ship, the Crystal Symphony, was to sail from Guam on February 2, 2020 over to the Philippines, Vietnam and Taiwan, disembarking in Hong Kong on Feb 15. "We looked forward to scuba diving in Saipan, visiting an old shipmate in Manila, and seeing the geological wonders of Ha Long Bay in Vietnam," they said. Ed was a speaker on the ship looking forward to sharing his love of the ocean and his concerns for its health.

The Sobeys started cruising when they first joined Semester at Sea as faculty and staff. Since that first voyage in 2008 they have taught and worked on three other full semesters and have branched out to speaking on commercial cruise lines. They have each traveled to more than 100 countries on six continents and Ed spent a winter in Antarctic giving him seven. Barbara is still trying to find a way to notch up her seventh continent. Between lecturing on cruise ships and training science teachers for Fulbright, the State Department, and universities, the couple travel four to six months a year.

The first they heard about the virus was a few days before their departure on the last day of January. So they were not surprised to learn that Crystal Cruises had changed the disembarkation port, but the company had not yet arranged a replacement for Hong Kong. They never thought about canceling, bolstered by their confidence that Crystal would look out for their best interests. Since Crystal had made the airline reservations, it was their responsibility to change them to the new disembarkation port.

First stop was meant to be Saipan in the Northern Mariana Islands but strong winds prevented them from docking. During the passage to the next stop, Manila, Crystal announced that the cruise would end in Taipei. That was fine with the Sobeys, as Crystal would change their return flights at no extra cost. But as they approached Manila, events took a twist. "We sailed into Manila on an overcast day quite ready to go ashore," said Ed. "As we passed Corregidor Island, the captain announced that the port of Manila was closed. We couldn't get off and neither could the dozens of crew who were scheduled to go home for their two or three month vacations."

The Crystal Symphony then headed to Vietnam, but it was only a matter of hours before the two remaining ports in northern Vietnam were closed. They then planned to dock in Ho Chi Ming City in the south. "This was good for us as the *World Odyssey*, Semester at Sea's ship, was docked there and we could visit shipmates and friends aboard her. I teach for Semester at Sea and had just left the ship December 23rd, 2019," said Ed. However, the port of Ho Chi Ming quickly closed, and Crystal Symphony had nowhere to go. "We bobbed around the South China Sea for several days. The Cruise Director was scrambling to fill the days. We speakers were busier than normal as were the entertainers," Ed described. "We could imagine what the home office must have been like. Trying to find a port that would accept us; re-booking airline tickets for passengers and crew; re-arranging contracts for the crew who were to come aboard or go home in Manila; ordering provisions and fuel in unexpected Singapore. Hats off to the band of people who were working around the clock."

The ship set up mandatory infrared temperature screening for passengers and crew. Twice a day everyone on board had to have their temperatures checked. As far as the Sobeys knew, no one on the ship became infected with the virus. "Many of our friends at home were emailing us with concerns for our safety. We, however, felt no hardship. We were cruising on a luxury ship with comfortable accommodations and first class service," said Barbara.

Eventually, the ship was permitted to land in Singapore. The ship had arrived a few days early so the Sobeys used it as their hotel while they wandered about. "First thing we noticed was that many locals were wearing face masks. Not uncommon throughout Asia, but the percentage of people with masks was higher than normal. Outside every mall pharmacy was a long line of people waiting to get in. The attractions, office buildings, and some malls had people taking temperatures of everyone passing."

The remainder of the journey was pretty smooth for the Sobeys, with Crystal taking care of their transportation to the airport and their flights home. "This story has two important messages for people interested in taking a cruise," said the Sobeys. "Go with the best cruise line you can afford. Crystal took care of us. You may have reasons to book your own flights but, if you have the cruise company book them, any hiccup in the schedule is their responsibility not yours. Some of our shipmates were spending hours and hundreds of dollars to change flights and then to change them again."

The couple's next cruise, from Tahiti to Fiji, was due to start in late March aboard Paul Gauguin Cruises. Ed's presentations on coral reefs, sharks, and island geology were all prepared, and they were both looking forward to diving at islands across the South Pacific. Of course that didn't happen, but they do want to travel by ship again. "We feel that the cruise ships won't risk bad publicity by sailing before it's safe. We restrict our cruises to smaller ships on the better lines. And, we are anxious to hear what extra precautions the industry is taking to make our next cruise safe."

Source: Personal communication with Ed and Barbara Sobey, May 2020

4

Introduction

As mentioned in Chapter 1, at the beginning of the COVID-19 outbreak, thousands of travelers – like our seafaring couple in the opening case study – had travel plans disrupted, and many were stranded abroad. This chapter will focus on such disruptions, but also touch on a certain segment of travelers who were oblivious to the crisis – either due to a lack of knowledge or to a lack of common sense! Consumer behavior (related to travel) during quarantine will then be examined, followed by a synopsis of the research undertaken during the lockdown concerning future travel behavior, as the industry sought to understand who would travel first and when, once lockdown regulations were eased.

Disruptions

Chapter 1 discussed how most governments introduced strict controls on overseas travel in response to the spread of COVID-19 leaving many tourists trapped abroad. Countries like Peru and Morocco put travel restrictions in place with little warning, essentially trapping some travelers for months (Mzezewe, 2020). Some countries were faster than others in bringing their nationals home and, even into May, tourists were still being repatriated. In Nepal, for example, more that 100 British travelers stranded in isolated parts of the country were rescued by the Gurkhas during the first week of May. Soldiers from the British Gurkhas Nepal network, along with UK embassy staff and drivers, traveled more than 4,000 miles through the Himalayas to reach tourists stuck in mountainous towns, villages and national parks, as part of a three-week rescue mission. British ambassador to Nepal Nicola Pollitt said: "Getting British nationals home in such an unprecedented time is a huge challenge around the world, but in a country like Nepal, with such extreme conditions, it would have been impossible to get everyone back without the close collaboration of the embassy and British Gurkhas Nepal" (Manning, 2020).

By early April, about 90% of destinations had completely or partially closed their borders for tourists (UNWTO, 2020). A significant increase was observed between March 9 and 24, when the number of destinations imposing travel restrictions more than doubled, from 81 to 181, following the declaration of COVID-19 as a pandemic by WHO on March 11, 2020. Figure 4.2 shows when, and to what extent, different measures were put in place across 209 international destinations. As of April 6th, nearly half of destinations worldwide (43%) had partially or completely closed their borders. Another 21% had introduced more selective, destination-specific, restrictions and 27% had suspended some or all international flights. The remaining 9% were relying on more varied visa measures, in-country travel restrictions, and/or quarantines.

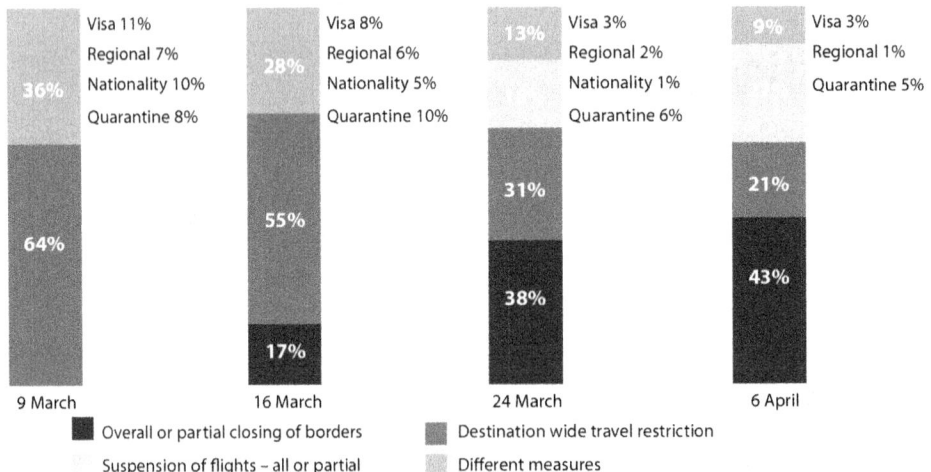

Figure 4.2: Changes in type of travel restrictions in place between March 9 and April 6 (courtesy of UNWTO)

Travel restrictions were swiftly followed by cancelations. One study in March asked more than 2,000 travelers from 28 countries about their travel behaviors during the pandemic. The results showed that 63.8% of the travelers planned to reduce their travel plans in the next 12 months. More than half had canceled their business travel immediately due to the coronavirus (Ali & Cobanoglu, 2020). Another study in the US found that 48% of leisure travelers had canceled their travel plans immediately (Longwoods International, 2020).

Research conducted by GlobalWebIndex early in March found that, in the UK, only 32% of vacationers (defined as those who usually take at least one vacation a year) hadn't had their travel plans affected by COVID-19. In the US, this percentage was just 8% (Beer, 2020). Almost half of US vacationers had canceled plans because they were forced to, and even more had done so of their own volition (see Figure 4.3).

Coronavirus has hit travel plans hard

% of U.S./UK vacationers who have done the following

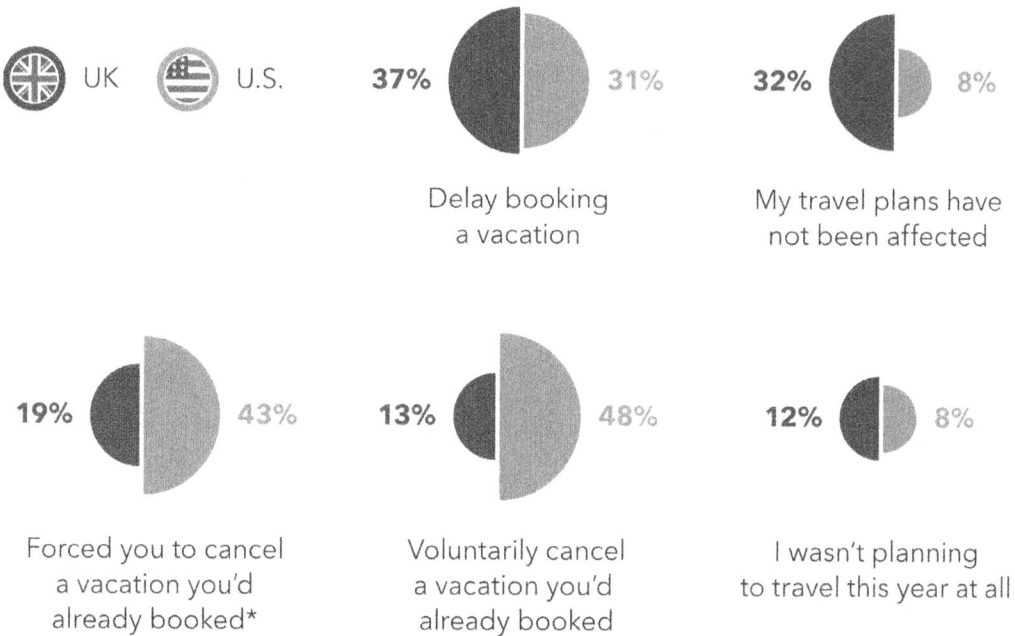

UK   U.S.

37%   31%   **32%**   8%

Delay booking
a vacation

My travel plans have
not been affected

19%   43%   **13%**   48%   **12%**   8%

Forced you to cancel
a vacation you'd
already booked*

Voluntarily cancel
a vacation you'd
already booked

I wasn't planning
to travel this year at all

Figure 4.3: The impact of COVID-19 on travel plans (courtesy of GlobalWebIndex)

The challenge for some businesses as travelers canceled their plans was that they simply could not afford to give refunds. At one point, the boss of Lufthansa Group issued a plea to passengers to be patient over cash refunds for canceled flights amid the COVID-19 crisis. Chairman and chief executive Carsten Spohr warned that the entire travel industry was at risk due to demands for refunds. With losses of €1 million an hour, almost all flights grounded and an appeal to governments for state aid, Spohr said: "We are fighting for the future of this company and the future of the roughly 138,000 employees of the Lufthansa Group. We are doing everything in our power to keep as many of them on board as possible" (Davies, 2020). Travel restrictions to prevent the spread of COVID-19 saw daily passenger carryings for Lufthansa deteriorate to 3,000 a day during lockdown from 350,000 before the crisis, with only 60 of the group's 760-strong fleet operating.

A pulse survey conducted by J.D. Power during the pandemic found that the travel industry was generally perceived by consumers to have comported itself well in reaction to cancelations. More than half of the travelers surveyed believed that travel suppliers were meeting or exceeding expectations on cancelation policies, and 60% of travelers agreed that hoteliers, cruise lines and airlines had demonstrated concern for the health and safety of the traveling public. However, the survey of 1,633 past-year business and/or leisure travelers conducted during the first comprehensive wave of travel cancelations, found that 33% of travelers thought the industry was not doing enough to help consumers protect themselves from illness while away from home. In general, the more often consumers traveled, the better they felt about the industry's response to the crisis.

Consumers canceling travel plans were particularly concerned about getting full refunds for prepaid (or non-cancelable) flights or hotel stays. "While the large hotel chains and airlines have focused on communicating messages of reassurance, the policies each sector has implemented are somewhat inconsistent and downright confusing. This is especially true for cancelation policies," said Andrea Stokes, practice lead for hospitality at J.D. Power (Christoff, 2020). J.D. Power noted that during times of crisis, companies which score high on customer satisfaction are usually the first to see their clients return, as pent-up demand drives travelers to return to the marketplace. Organizations like Micato (profiled in Chapter 3) and Canadian airline WestJet (from the author's experience) understand this. Both have a reputation for being customer-focused (Hudson & Hudson, 2017), and made processing cancelations and refunds a priority during the crisis.

For travel companies dealing with cancelations, these were challenging times. Omio, for example, a travel search engine for the UK & Europe, witnessed a big spike in demand for customer service as worried customers looked to cancel trips. "The whole company is actually stepping in to help customer service because we've seen a spike in cancelations," said CEO Naren Shaam. For TravelPerk, which focuses on corporate travel, it was a similar story. The company saw a dramatic decrease in bookings, and had to move some of its sales team to customer support to deal with disruptions. TravelPerk also noted radical changes to the usual booking window. "Most of the trips we see right now is somebody booking for tomorrow or for two days from now because they know they can travel or have certainty they can travel," said CEO Avi Meir in March. "This is unusual compared to normal times. In normal times people book 20-21 days ahead on average. So you have a huge decrease in the booking window" (Lomas, 2020).

Figure 4.4: Travelers were urged to be patient waiting for refunds (photo by Erik Odiin on Unsplash

For those in the travel insurance industry, these were also testing times. Travel insurance providers were finding it difficult to deliver on pledges to consumers, like those made by the Association of British Insurers' (ABI) travel insurance members. Two of these pledges include considering "all valid travel insurance claims quickly and fairly" and implementing "business continuity plans to be able to continue to handle travel insurance claims in challenging circumstances" (Grzadkowska, 2020). COVID-19 was set to cost UK travel insurers at least $335 million and see them record their biggest annual loss ever, according to ABI. The vast majority of losses were for cancelations, with some for disruption costs incurred overseas (Norris, 2020).

During the pandemic, many travelers were confused over insurance and voiced their concerns. Navigating the 'will it or will it not pay out' landscape was a challenge, with companies issuing statements like: "COVID-19 related illness is covered unless your destination has a Level 3 (avoid non-essential travel) or 4 travel advisory (avoid all travel) in place prior to your departure". In the US, a House Oversight subcommittee demanded that travel insurance providers explained why they were not fully covering trip cancelations stemming from the pandemic (Grzadkowska, 2020). The varied approach of each travel-related company did not help, with some like Eurostar or British Airways offering e-vouchers for trips, while other companies offered full refunds, even for non-refundable bookings.

Travel insurance comparison site Squaremouth noted that when the health emergency was elevated to the status of a 'pandemic', travel insurance coverage for coronavirus was limited. They were recommending that travelers purchase a 'cancel for any reason' policy, which has its own limitations as a time-sensitive and more expensive upgrade. Many major insurers were saying they would no longer cover emergency medical expenses related to COVID-19 for travelers visiting countries with high-level travel advisories. Travel insurance usually helps people who have to cancel trips, or receive medical care abroad, due to unforeseen events. But as the coronavirus pandemic continued, insurance companies were deeming the virus a 'known event' that customers were aware of before leaving the country (O'Hara, 2020).

Moving forward, Alex Sharp, Managing Director for UK-based travel insurance specialist ASUA Group, believes that because of the huge number of claims in the pipeline, not every business will be able to survive. "It is both the scale and duration of this crisis that makes it almost impossible for individual insurers – or the market as a whole – to mitigate people's losses and disruption ... claims will take a hugely long time to settle as the travel trade struggles to work out exactly what costs can be recovered and what is lost," he said (Bullock, 2020). Kasara Barto, PR Manager at Squaremouth, sees the pandemic as a major shift in the way that travel insurers will operate in the future. "Providers will include coverage for virus outbreaks on future policies, similar to the 9/11 attack leading to the addition of the Terrorism benefit. Prior to 9/11, the Terrorism benefit was not offered on travel insurance policies. However, upon the outpouring of traveler concern after the attack, many providers added it to their standard coverage," she said (Bullock, 2020).

The oblivious

Not all tourists saw a disruption in their travel plans. Some were oblivious to the COVID-19 outbreak simply because they were off the grid. A group of a dozen or so rafters, for example, were on a 25-day adventure down the Colorado River through the Grand Canyon during the outbreak. When they left mid-February, cases of the coronavirus were showing signs of decline in mainland China, so as they took the last paddle strokes of their journey on March 14, they were completely unaware and shocked to find that the virus had exploded into a once-in-a-generation pandemic, setting off a global health and economic crisis and shutting down large parts of American life. One of the rafters, Mason Thomas, said he had been monitoring the spread of the virus before he left. "The last thing I thought before I got into the canyon was: 'Well, I can't do anything about it. Maybe it won't be there when I get back.' I figured if it was bad maybe we'd just all be washing our hands even more or something more serious. But not this" (Warzel, 2020).

Australian academic Dr. Christiaan De Beukelaer was also unaware of the outbreak for weeks as he sailed across the Atlantic, studying life on a cargo sailboat, the *Avontuur*. But on March 18, when the ship was halfway across the ocean, the ship's owner, Cornelius Bockerman sent a message to the captain on the satellite phone: "Dramatic times! You will not find the world the way it was before you set sail." From the ship, De Beukelaer explained: "We're so disconnected from the world. It's almost as if the whole pandemic is something that is happening like in a distant galaxy somewhere" (Griffin, 2020).

Other travelers seemed to be oblivious to the COVID-19 outbreak through their actions. Students on spring break in Florida, for example, earned fierce criticism for vacationing as normal in Miami as the virus spread through America. "If I get corona, I get corona," one student said. "At the end of the day, I'm not going to let it stop me from partying. I've been waiting, we've been waiting, for Miami spring break for a while. About two months we've had this trip planned, two, three months, and we're just out here having a good time" (Brito, 2020). Tourists, though, were receiving mixed messages from Florida's leaders at the time. The Ultra Music Festival, a marquee electronic dance music event that draws tens thousands of people, was canceled, whereas the Winter Party Festival, a beachside dance party and fundraiser for the LGBTQ went ahead.

Figure 4.5: Miami Beach received some criticism for allowing the Winter Party Festival to go ahead early in March (photo by Marc Fanelli-Isla on Unsplash)

The exact number of people who returned from vacation trips to Florida with the virus may never be known, but cases as far away as California and Massachusetts have been linked to the Winter Party Festival. Mayor Francis Suarez of Miami, one of the first elected officials in the country to test positive for the coronavirus, said other jurisdictions' decisions to keep events going proved costly. "That ended up as a national embarrassment, when you saw what happened with the spring breakers and what happened unfortunately, tragically, with the music festival," he said (Mazzei & Robles, 2020).

Quarantined

As mentioned in Chapter 1, Italy was the first country in European to put some of its citizens into quarantine and, ironically, the very word "quarantine" has Italian roots. In 1374, in an effort to protect coastal cities from the Black Death ravaging 14th-century Europe, ships arriving in Venice from infected ports were required to sit at anchor for 40 days (*quaranta giorni*) before landing, a practice that eventually became known as quarantine – derived from 'quarantino', the Italian word for a 40-day period (Vuković, 2020). Other countries soon imposed quarantines. In 1377, Ragusa (present-day Dubrovnik, Croatia), passed a ground-breaking law to prevent the spread of the pandemic requiring all incoming ships and trade caravans arriving from infected areas to submit to 30 days of isolation. The legislation, *Veniens de locis pestiferis non intret Ragusium vel districtum* ("Those arriving from plague-infected areas shall not enter Ragusa or its district"), stipulated that anyone coming from pestiferous places must spend a month in the nearby town of Cavtat or the island of Mrkan for the purpose of disinfection before entering the medieval walled city. Dubrovnik's quarantine measures were seen as more just and less discriminatory than Venice's and so prevailed around the world.

Quarantine measures introduced for travelers (or nationals returning home) during the COVID-19 pandemic were not as restrictive as those in the 14th century, but were nevertheless a deterrent to travel. Austria, for example, imposed a 14-day quarantine period for incoming visitors during the pandemic (the 14-day period seemed to the be norm in other countries), although those wanting to avoid the quarantine in Austria could pay $200 for a test to prove they were free of COVID-19. Medical experts encouraged people to isolate for 14 days because one could be contagious for up to two weeks after being exposed to COVID-19. Some workers in the travel sector were forced into isolation – even into May 2020 more than 90,000 cruise ship workers were still stranded on ships, many of them confined to their small cabins for 21 hours a day (Levin & Oanh Ha, 2020).

As mentioned in Chapter 2, as the world went into lockdown, businesses had to pivot and look for ways to adapt. A creative example of this 'COVID-aptability' was a restaurant in Amsterdam that came up with the idea to allow people to enjoy their favorite meal along with effective implementation of social distancing. The restaurant, by the name of Mediamatic ETEN, built small quarantine greenhouses where people could sit and enjoy dinner by the canal. As it waited for permission from the government to re-open, the restaurant tested the candle-lit, four-course vegetarian experience by offering limited service to family and friends of staff. The restaurant is part of Mediamatic, an art center dedicated to new developments in the arts. The center organizes lectures, workshops and art projects, focusing on nature, biotechnology and art and science. Other organizations decided that if customers couldn't get to them, they would go to the customers. Destinations like the Faroe Islands saw the quarantine period as an opportunity to allow consumers to experience their products and experiences virtually.

Figure 4.6: Social distancing dining at Mediamatic ETEN in Amsterdam (courtesy of Mediamatic, photo by Anne Lakeman and Willem Velthoven)

The UK similarly brought a 'little bit of Britain' to those in lockdown with their campaign *Showing love for Great Britain* (see end of chapter case study). Californian wineries held virtual tastings amid the pandemic giving people the wine country experience from the comfort of their own homes. Those interested could purchase the tasking kits online and then join the virtual tours to replicate the guided tastings. In Hong Kong, after a series of major art events were canceled in March – including Art Basel in Hong Kong, arguably Asia's most important art fair – more than 80 galleries, museums, auction houses, not-for-profit organizations, educational institutions and cultural media outlets joined hands to launch Art Power HK, a campaign and free online platform promoting everything that was happening in Hong Kong's art scene. The website featured video interviews with artists, video tours of exhibitions in Hong Kong, and a series of live panel discussions (Lo & Giles, 2020).

In the events sector, Formula 1 launched a virtual Grand Prix series during lockdown – live streaming with real drivers competing against each other – just for entertainment. The broadcast was available on the official Formula 1 YouTube, Twitch and Facebook channels, as well as F1.com, and attracted an impressive 350,000 views over its hour-and-a-half run time. Julian Tan, Head of Digital Business Initiatives and Esports said: "We are very pleased to be able to bring some light relief in the form of the F1 Esports Virtual GP, in these unpredictable times, as we hope to entertain fans missing the regular sporting action. With every major sports league in the world unable to compete, it is a great time to highlight the benefits of esports and the incredible skill that's on show" (Stoddart, 2020).

Burning Man was another popular event to go online. Held annually since 1986, Burning Man usually attracts tens of thousands of people who gather in Nevada's Black Rock Desert to create Black Rock City, a temporary metropolis dedicated to community, art, self-expression, and self-reliance. Announcing the virtual event, planners said they would "lean into" the extravaganza's previously announced "multiverse" theme by re-creating its desert culture in cyberspace. Organizers tried to persuade ticket-holders not to request refunds. "Some of you who already purchased a ticket for the playa may need that money now more than ever," they said. "We're committed to providing refunds to those who need them, but we're also committed to keeping Burning Man culture alive and thriving, and to ensuring our organization stays operational into next year's event season – which will require substantial staff layoffs, pay reductions, and other belt-tightening measures. Burning Man Project's survival is going to depend on ingenuity and generosity. Luckily, our community is rich in both" (Burning Man Project, 2020).

Burning Man Project
@burningman

In the interest of the health & wellbeing of our community, we have decided not to build Black Rock City this year. Burning Man, however, is alive & well, and we look forward to seeing you in the Multiverse. Read more in the Burning Man Journal.

The Burning Man Multiverse in 2020
After much listening, discussion, and careful consideration, we have made the difficult decision not to ...
journal.burningman.org

7:01 PM · Apr 10, 2020 · Twitter Web App

Figure 4.7: Burning Man 2020 went online for the first time in its history

Smaller events also moved online. For example, in England, Dorset's annual knob-eating competition was held online for the first time. The event – in which contestants vie to eat more of the county's traditional biscuits than their rivals – usually draws huge crowds. But instead, the 100 competitive eaters live-streamed their attempts to swallow the savory spheres. The winner was local Kate Scott who ate eight and a half of the biscuits in one minute. The dry savory biscuits have been made by Moores of Morecombelake for more than 150 years. Their name comes from the hand-sewn Dorset knob buttons that were also made locally. The knobs are baked three times and due to their hardness, eaten after first being soaked in sweet tea. They are traditionally accompanied by Dorset Blue Vinney cheese and were said to have been a favorite food of local author Thomas Hardy (Hall, 2020).

Understanding future traveler behavior

During the pandemic, researchers sought to understand traveler sentiment in order to gauge future travel behavior. Misty Belles, Managing Director of global public relations at Virtuoso, an international travel agency network with 1,100 locations, said that their clients were falling into one of three groups: "About one-third are canceling outright, a third are postponing and a third are in 'wait-and-see' mode" (Fletcher, 2020). Clayton Reid, CEO of MMGY Global suggested at the end of March that there would be four phases of societal mindset during the pandemic; fear, understanding, action, and rational behavior. "Today we are still in the fear phase, which remains to play out as we near infection spikes globally, but understanding and action phases are now in nascent stages and rational behavior will follow soon after that," he said (Reid, 2020). His belief was that travel sentiment, activity and spending would return more quickly than many had predicted. A study about the same time by the Brandon Agency (2020) found that a third of vacationers could not be persuaded to book travel at all during the coronavirus outbreak. But a strong travel insurance policy (26%) and flight discounts (19%) would attract *some* travelers to book during this time.

How Much Do You Miss the Following?
(% Saying "Miss Terribly" or "Miss A Lot")

Vacations (68.4%)

Dining in Restaurants (59.9%)

Planning Travel (56.1%)

Weekend Getaways (52.8%)

Live Concert/Musical Performance (35.1%)

Going to Museums (33.4%)

Going to Bars (32.9%)

Attend Professional Sporting Event (30.5%)

Business Trip (19.3%)

Going to Conventions (16.9%)

Destination ◆ Analysts

Figure 4.8: What travelers were missing during lockdown (courtesy of Destination Analytics)

Influencer Marketing (2020) said there was a decrease of nearly 50% in traffic to travel websites during lockdown, but research by Destination Analytics (2020) suggested that over half of consumers in the US were missing the very act of planning travel. A further 68.4% of people were yearning for a vacation, and 60% of consumers were craving restaurants (see Figure 4.8). Destination Analytics also asked American consumers early in May what channels they would be most receptive to in terms of learning about travel destinations (see Figure 4.9). Over 85% of Millennial and GenZ travelers – and 7 in 10 GenX and Boomer travelers – cite a digital resource as the forum through which they would be most receptive to travel messaging, with social media platforms like Instagram and Facebook as well as search engine marketing appearing to be the most popular. Email was also one of the best ways to reach all ages of travelers who were in a state of openness to travel messaging. Travelers were looking for destinations to speak to them in an honest (59%) but friendly (40%) tone in advertising.

Where Are You Most Receptive to Learning about Travel Destinations Right Now?

	Millennial/GenZ	GenX	Baby Boomers
Instagram	32.7%	12.9%	5.8%
Facebook	25.5%	20.7%	15.2%
Websites found via Search Engine	20.4%	33.5%	40.1%
Online Articles/Blogs	18.6%	19.6%	20.7%
TikTok	16.6%	2.0%	0.8%
Ads around the Internet	16.6%	16.5%	19.0%
Email	16.1%	25.8%	30.6%
Twitter	15.9%	6.3%	3.3%
Pinterest	15.4%	5.5%	4.5%
Text Messages	13.7%	7.3%	3.2%
Digital Influencers	8.4%	4.2%	1.0%
Apps	7.6%	4.3%	2.7%
None of these	14.1%	29.8%	25.8%

Destination ◆ Analysts
DO YOUR RESEARCH

Figure 4.9: Media channels US travelers were most receptive to during the crisis (courtesy of Destination Analytics)

Prior research has indicated that some travelers are more likely to travel than others during and after a crisis. Hajibaba et al. (2015) were the first to propose that a segment of tourists exists which is inherently more resistant to crises than other tourists. They found in their study that this market is younger than average, more extrovert, willing to take high physical risks, motivated to travel by opportunities related to sports and health, and actively engaged in activities such as mountain biking, horse riding and hiking. Understanding this segment likely means identifying which are the first tourists to go to a destination either during or after a crisis event – essential information for crisis management planning. Hajibaba et al. suggest that these tourists are an attractive market segment for travel providers, intermediaries and destinations, not only because of their crisis-resistance, but also because they are highly targetable. They engage in very specific activities at the destination, and they are also highly involved in the travel planning process; therefore, they can be influenced directly through a variety of channels, including social media.

As lockdown restrictions eased, travel marketers tried to persuade travelers to book while the prices were low. Jesse Neugarten, CEO and founder of the million-plus member discount subscription service Dollar Flight Club said in April: "What we're telling our customers is that they book flights now for future travel, because prices are going to be as cheap as you've seen since 2001, and then they're going to increase sharply when demand rebounds." He also pointed out that there were less risks in booking travel. "Now, the majority of major US carriers are offering free change fees and cancellations," he said (Fletcher, 2020). Hotels, too, loosened up their relatively fair cancelation and rebooking policies. Arne Sorenson, President and CEO of Marriott International, wrote to customers on April 8: "For guests making new reservations for any future arrival date, including reservations with pre-paid rates, between March 13 and June 30, 2020, we will allow the reservation to be changed or canceled at no charge up to 24 hours before your scheduled arrival date." Pre-paid rates are usually discounted and nonrefundable.

At the University of Florida, researchers at the Tourism Crisis Management Initiative conducted an extensive and continuous study on COVID-19 travel risk perceptions during the crisis. They used up-to-date data to understand American travelers' perceptions of risk towards travel as well as other variables related to COVID-19. As lockdown restrictions eased in May, travelers were less anxious about traveling than in the previous two months, but they were looking for safety reassurances from travel providers. Figure 4.10 shows the operational practices travelers expected to see in place if they were to visit attractions, accommodations or airports.

Research by Globalwebindex during the lockdown period found that about 80% of consumers expected to change their travel behaviors in some way in the future (Mander, 2020). Around 3 in 10 said they would be taking more staycations, a similar proportion would take more domestic rather than foreign vacations, while around a quarter expected to take fewer vacations. When asked what would give them the confidence to start booking trips again, by far the most popular answer was "when I feel it's safe to travel" (58%). Vacationing in the local area or in one's own country was very clearly part of the reaction to this. Those planning more domestic trips (68%) or staycations (65%) were the most likely

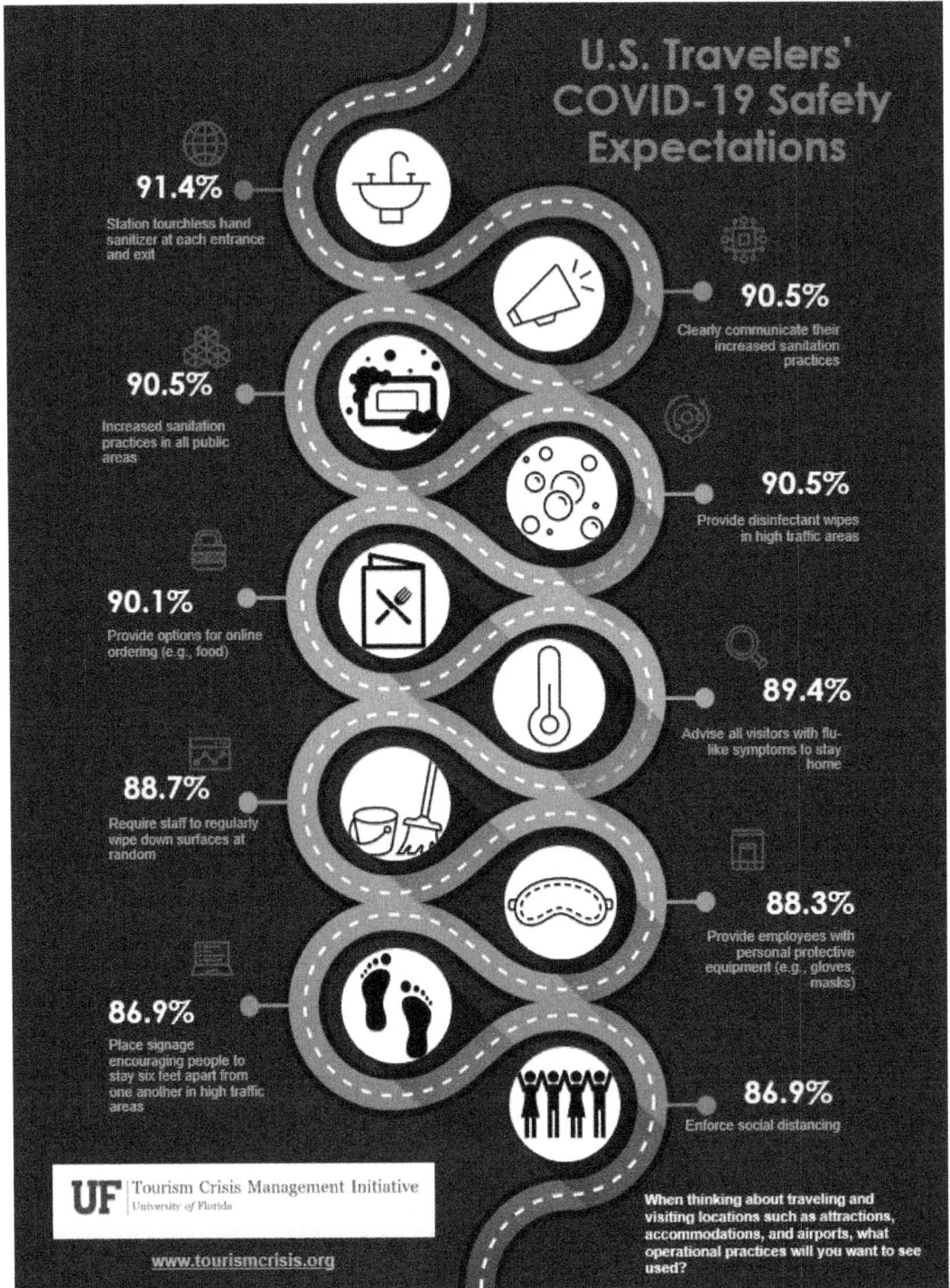

U.S. Travelers' COVID-19 Safety Expectations

91.4%
Station touchless hand sanitizer at each entrance and exit

90.5%
Clearly communicate their increased sanitation practices

90.5%
Increased sanitation practices in all public areas

90.5%
Provide disinfectant wipes in high traffic areas

90.1%
Provide options for online ordering (e.g., food)

89.4%
Advise all visitors with flu-like symptoms to stay home

88.7%
Require staff to regularly wipe down surfaces at random

88.3%
Provide employees with personal protective equipment (e.g., gloves, masks)

86.9%
Place signage encouraging people to stay six feet apart from one another in high traffic areas

86.9%
Enforce social distancing

UF | Tourism Crisis Management Initiative
University of Florida

www.tourismcrisis.org

When thinking about traveling and visiting locations such as attractions, accommodations, and airports, what operational practices will you want to see used?

Figure 4.10: US travelers' COVID-10 safety expectations (courtesy of the Tourism Crisis Management Initiative)

to say it was a feeling of safety that would prompt them to start booking again. Consumers were also planning a reduction in their leisure behaviors once the crisis was over. Figure 4.11 shows that 29% of consumers planned to visit bars and pubs less often although, interestingly, for regular alcohol drinkers, this rises to 33%. For fast-food restaurants, 31% overall planned to eat at such outlets less frequently; among regulars, this ticks up to 35%.

Planned reduction in leisure behaviors

% who say they expect to do the following after the outbreak is over

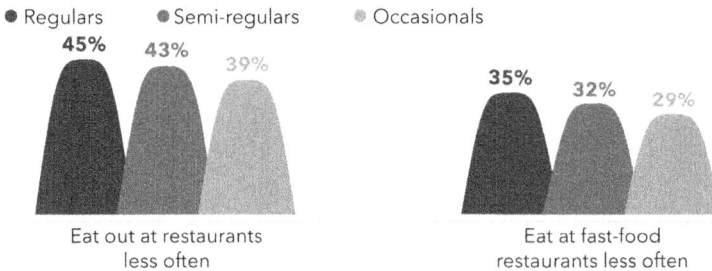

● Regulars   ● Semi-regulars   ● Occasionals

45%   43%   39%

35%   32%   29%

Eat out at restaurants
less often

Eat at fast-food
restaurants less often

Question: After the outbreak is over, do you think you'll do any of the following? Source: GlobalWebIndex Custom Research, April 22-27 2020 Base: Internet Users aged 16-64 in 17 countries

Figure 4.11: Planned reduction in leisure behaviors after the crisis (courtesy of GlobalWebIndex)

In order to counter this reluctance to eat and drink outside the home after the pandemic, Vienna handed out 50-euro vouchers to every family in the Austrian capital to spend in local restaurants and cafes after they reopened in May. Mayor Michael Ludwig said that all 950,000 households in Vienna would receive vouchers to help restaurants and cafés get back on their feet after a two-month coronavirus lockdown (Groendahl, 2020).

Figure 4.12: Vienna in Austria, where families were given vouchers to encourage them to go to restaurants and cafés after the lockdown (photo by Levy M on Unsplash)

Finally, from a study of US travelers in early May, Destination Analytics (2020) found that the vast majority would approach travel with trepidation as they thought about starting again (see Figure 4.13). McKinsey (2020) found similar hesitation amongst consumers elsewhere in the world, with the majority saying they did not intend to undertake international travel soon, while travelers in several countries – with the exception of Germany and France – planned to restrict domestic travel as well.

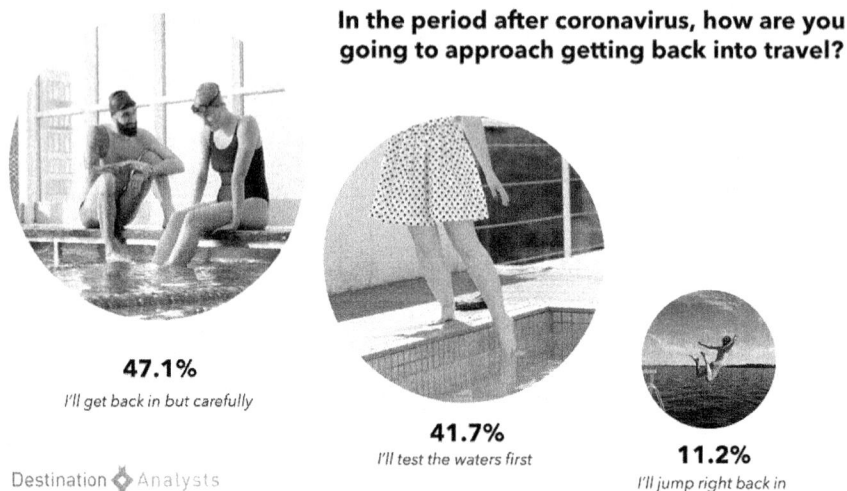

In the period after coronavirus, how are you going to approach getting back into travel?

47.1%
I'll get back in but carefully

41.7%
I'll test the waters first

11.2%
I'll jump right back in

Destination ◆ Analysts

Figure 4.13: Perceptions about travel as lockdown restrictions were eased (courtesy of Destination Analytics)

A halo effect

The halo effect is a type of cognitive bias, and is the tendency for positive impressions of a person, company, brand or product in one area, to positively influence one's opinion or feelings in other areas. Previous research has shown that the perceived image of a country in general can have a significant influence on how people develop beliefs about the country as a travel destination (Lee & Lockshin, 2011). As countries eased lockdown restrictions, some tourism destinations were hoping that there might be a halo effect resulting from the positive media they had received in dealing the crisis and for being perceived as 'COVID-free'.

Portugal, for example, was planning to take advantage of being perceived as a 'safe' destination. Elidérico Viegas, President of the Algarve Hotelier Association (AHETA), said that the region was setting itself apart from rival destinations such as Spain and Italy, where the outbreak has had devastating consequences. "The pandemic has had a relatively low impact in Portugal, and in the Algarve in particular," said Viegas. "Portuguese and foreign holidaymakers, particularly British, are looking to us as a COVID-safe destination. Hotels are being contacted by an increasing number of people looking to book their summer holidays here. For example, the British and German governments have advised their citizens against traveling to Spain. This could be a good opportunity for the Algarve. Around 18 million British and 11 million German holidaymakers travel to Spain, just to give you an idea of the impact these markets have on the sector" (Bruxo, 2020).

South Pacific island nations were also receiving attention for having dodged the worst ravages of the pandemic. Even into May, around a dozen Pacific island nations remained virus-free, most of them remote dots in the ocean that sealed their borders when they saw the carnage COVID-19 was causing elsewhere in the world. But the economic impact of the pandemic has been devastating. Communities across the region rely heavily on tourism – in some places as much as 50% of GDP – that comes from parts of the world where COVID-19 has been actively spreading. However, an idea to include the islands in a quarantine-free travel 'bubble' with Australia and New Zealand – where infection rates were also low – was getting only a cautious welcome. "We believe that little pockets like ourselves, currently free of the virus and working with other like-minded countries in the region, exercising caution, should be able to reopen," said Cook Islands Tourism chief executive Halatoa Fua. But he stressed the bubble would need to come with stringent safeguards. Palau's tourism minister F. Umiich Sengebau said in the absence of direct air links with Australia and New Zealand, it made more sense for his country to pursue a bubble arrangement with Taiwan. Palau is one of the few nations to have diplomatic relations with Taipei. "This is an ingenious idea that we must consider for a country like Taiwan, which has done a very good job in handling of COVID-19 pandemic," he said (Bangkok Post, 2020).

Some Asian countries with a growing reliance on tourism may also benefit from this halo effect. Vietnam, for example, had recorded only around 300 cases of COVID-19 and not a single death by mid-May, and had been lauded for its approach to controlling the virus. The country chose prevention early and, on a massive scale, enacting measures other countries would take months to move on (see case study in Chapter 6). By early January, before it had any confirmed cases, Vietnam's government was initiating an action plan to prepare for the virus, and when the first virus case was confirmed on January 23 – a man who had traveled from Wuhan to visit his son in Ho Chi Minh City – Vietnam's emergency plan was in action. Vietnam's tourism industry was significantly affected by the pandemic, with practically no international arrivals for three months. However, while the situation is unprecedented, tourism experts in the country expect the industry to rebound faster and stronger than competitors. Vietnam relies heavily on Chinese and South Korean tourists, which accounted for 56% of its international arrivals in 2019. This also presents an opportunity, as China and South Korea have also largely contained the pandemic (Samuel, 2020).

During the lockdown, those who had to cancel or cut short trips to Vietnam, or who were unable to travel to the country, could enjoy exploring Vietnam with a 'Stay at Home' kit, created by the Vietnam National Administration of Tourism

(VNAT) together with the Vietnam Tourism Advisory Board (TAB). The kit, free on www.vietnam.travel featured 360-degree interactive tours of Vietnam's renowned UNESCO World Heritage Sites, easy recipes for popular Vietnamese dishes, and suggestions for cultural explorations in books, music and art. Would-be visitors could also download a virtual background of popular Vietnamese destinations for their Zoom meetings (see Figure 4.14).

Figure 4.14: One of the Zoom backgrounds offered by Vietnam as part of their 'Stay at Home' kit (courtesy of Vietnam Travel).

Finally, Iceland was hoping to capitalize on its handling of the pandemic and reopen to tourism by mid-June, offering free COVID-19 tests on arrival. Those who tested negative would be at liberty to enjoy their time in the country, but those who tested positive would have to self-isolate for 14 days. "Iceland's strategy of large-scale testing, tracing, and isolating have proven effective so far. We want to build on that experience of creating a safe place for those who want a change of scenery after what has been a tough spring for all of us," said Thordis Kolbrun Reykfjord Gylfadottir, Minister of Tourism, Industry and Innovation (Hosie, 2020). With an economy that is hugely reliant on tourism, the island nation was eager to find ways to reopen safely to the rest of the world. Gudlaugur Thór Thórdarson, Iceland's Minister for Foreign Affairs, said in May: "Although Iceland is an island, it has always thrived through international trade and cooperation. With only three cases of the virus diagnosed in May, we are once again ready to carefully open our doors to the world. While we remain cautious, we are optimistic as a country that we can successfully begin our journey back to normality" (Government of Iceland, 2020).

Case study: Showing love for Great Britain

There was widespread optimism that 2020 would bring be a bonanza for British tourism, an industry worth £127billion annually to the UK economy. About 40 million people were expected to flood into the UK to take advantage of attractions in hundreds of historic towns and cities – the highest figure ever. Then came the pandemic, and the UK's DMO, VisitBritain, has now forecast a 54% drop in international visits for the year, leading to a 55% fall in overseas' visitor spending. "This year has seen domestic and global travel brought to a standstill. While assessments of the impact of this global crisis on tourism are quickly surpassed by the fast-changing reality, there is no doubt the industry has been one of the earliest and hardest hit of all economic sectors," said VisitBritain chief executive Patricia Yates.

In the light of COVID-19, VisitBritain completely refocused its international marketing activity. Unable to promote travel to Britain during lockdown, the DMO decided to maintained a dialogue with consumers to ensure Britain remained top of mind when international travel started back up. The plan was to bring a little bit of Britain to international visitors facing coronavirus travel bans. The campaign, *Showing love for Great Britain*, promoted popular British culture from literature, film and music to heritage, gardens and food and drink across social channels, with the sharing of inspirational content such as recipes, playlists and even pub quizzes for visitors to enjoy from home.

Love GREAT Britain
March 18 ·

We know it is a difficult time and we can't welcome you into our country with open arms to enjoy our afternoon teas, our pints of ale, singing on the terraces, walks in our beautiful countryside, rambles around stately homes and beautiful gardens in bloom.

So, for now, we're going to be bringing a bit of Britain to you.

We'll be sharing British films and TV series to binge, some excellent British playlists for you to rock out to in your living room and recipes from some of Br... See More

1.2K   102 Comments 135 Shares

Figure 4.15: A Facebook post from VisitBritain explains the showing love campaign

As an example, followers were invited to explore 'virtually' the filming locations featured in the Harry Potter films. Tweets from @VisitBritain provided information about a dozen Harry Potter film locations, from the Glenfinnan Viaduct in Scotland where the Hogwarts Express traveled on the route to Hogwarts School of Witchcraft and Wizardry, to Alnwick Castle in Northumberland, scene of Madame Hooch's broomstick flying lesson in the first film. The phenomenon of Harry Potter – or 'Pottermania' – was leveraged previously by VisitBritain to rescue the country's ailing tourism industry hit by the terrorist attacks of 2001. At that time, in an integrated marketing strategy, VisitBritain utilized the books and the films for their promotional potential, increasing tourist visits through-out Britain and, in particular, to destinations featured in the books and movies.

VisitBritain 🇬🇧 ✔
@VisitBritain

Looks like now's the perfect time to re-watch all eight Harry Potter films... 🧹

See how many of these filming locations from across the UK you can recognise. 🎬 ➡ ow.ly/30R050yP1sl

#LoveGreatBritain

Figure 4.16: A tweet from VisitBritain on March 19, 2020 promoting the film locations for the Harry Potter movies

VisitBritain also capitalized on the UK's rich musical heritage by creating dedicated Spotify playlists for followers and by promoting famous music locations in the country. These two tweets, for example, offered a playlist of music that came out Liverpool, and a link to more information about the city's musical locations, such as the Cavern Club, the birthplace of Britain's most iconic band, the Beatles. The campaign also featured musical locations in Manchester and London. Bands like The Smiths, Oasis and Take That have their roots in Manchester, and the 'London's Calling' Spotify play-list featured artists like David Bowie, The Clash, Elton John and Queen.

VisitBritain 🇬🇧 ✔ @VisitBritain · Apr 19

With bands coming out of Liverpool including Frankie Goes to Hollywood, Echo & the Bunny Men and oh yes, The Beatles, we had a lot of fun putting together this dedicated playlist 🎧 ➡
open.spotify.com/playlist/0gLHf...

#LoveGreatBritain

Liverpool and Merseyside Sounds - The Beatles, The Searchers, Billy Fury, and others
🔗 open.spotify.com

💬 2   🔁 9   ♡ 17   ↥

VisitBritain 🇬🇧 ✔ @VisitBritain · Apr 19

Learn more about the musical locations in Liverpool that have seen history made and have given Liverpool such a unique energy ➡
visitbritain.com/gb/en/liverpoo...

#LoveGreatBritain

Get into the groove with Liverpool's musical high...
Discover the best Liverpool's music scene has to offer, from Beatles tributes to the latest and ...
🔗 visitbritain.com

💬   🔁 2   ♡ 6   ↥

Figure 4.17: Tweets from VisitBritain on April 19, 2020 promote musical locations in Liverpool

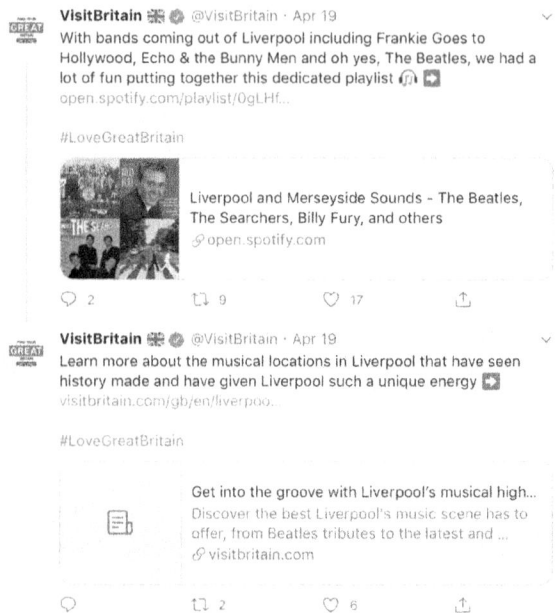

VisitBritain also promoted a number of virtual tours of the UK's popular attractions encouraging tourists to "try before you buy". These virtual tours allowed the potential visitor to explore the attraction via navigation buttons that put them in complete charge of their virtual viewing. Some were also accompanied by a soundtrack recorded on site, to add additional atmosphere to the experience – for example, virtual tourists taking in the sites of Conwy Castle could hear the sounds of the seabirds overhead and the chiming of the clocktower as they explored. Other virtual tours featured during the campaign were the Palace of Holyrood House in Edinburgh, the Tower of London, the Palace of Westminster, and the Giant's Causeway.

Media around the world picked up on VisitBritain's *Showing love for Great Britain* campaign. An article in the *Toronto Sun* in Canada, for example, featured a number of traditional British recipes that had been part of the social media campaign. Presenting the recipe for the British scone, the author said: "What everyone loves to celebrate is the typically British moment in the afternoon — afternoon tea! It's a beloved ritual enjoyed around the world, but when you celebrate in its traditional home, it's beyond wonderful. Originating in the southwest of England, in Cornwall and Devon, the tradition has spread throughout Great Britain. All you need in addition to a hot kettle of tea is some clotted cream or curd (a very thick cream), some tasty preserves ... and, of course, scones!"

References: DeMontis (2020). VisitBritain.com

4

References

Ali, F. & Cobanoglu, C. (2020). Global tourism industry may shrink by more than 50% due to the pandemic. *The Conversation*, 21 April. https://theconversation.com/global-tourism-industry-may-shrink-by-more-than-50-due-to-the-pandemic-134306

Bangkok Post (2020). Virus-free islands weigh risks of re-opening to tourists. *Bangkok Post*, 8 May. https://www.bangkokpost.com/world/1914836/virus-free-pacific-islands-weigh-risks-of-re-opening-to-tourists

Beer, C. (2020). How travel is being affected by the COVID-19 outbreak. *GlobalWebIndex*, 17 March. https://blog.globalwebindex.com/chart-of-the-week/travel-in-the-time-of-coronavirus/

Brito, C. (2020). Spring breakers say coronavirus pandemic won't stop them from partying. *CBS News*, 25 March. https://www.cbsnews.com/news/spring-break-party-coronavirus-pandemic-miami-beaches/

Bruxo, M. (2020). Summer bookings pick up pace: tourists look to Algarve as "Covid-safe" destination. *Portugal Resident*, 28 April. https://www.portugalresident.com/summer-bookings-pick-up-pace-tourists-look-to-algarve-as-covid-safe-destination/?fbclid=IwAR2jT2hq4yrYI1X1qpJygIvOGMrooGomxlwvDKy-nKtUxugNzxgWP4flt-w

Bullock, C. (2020). The global travel insurance industry adapts to Covid-19. *ITIJ*, 4 May. https://www.itij.com/latest/long-read/global-travel-insurance-industry-adapts-covid-19

Burning Man Project (2020). The Burning Man multiverse in 2020. *Burning Man Journal*, 10 April. https://journal.burningman.org/2020/04/news/official-announcements/brc-2020-update/

Christoff, J. (2020). Survey shows travelers satisfied with industry response to outbreak. *Travel Pulse*, 27 March. https://www.travelpulse.com/news/features/survey-shows-travelers-satisfied-with-industry-response-to-outbreak.html

Davies, P. (2020). Cash refund demands 'endangering entire travel industry' – Lufthansa boss. *Travel Weekly*, 6 May. https://www.travelweekly.co.uk/articles/370346/cash-refund-dema...=IwAR3-slQleSNTPQHFg9D82CZl85UQZqkzbghcRq2HXcTiDu_6vyDoqMfytdQ

DeMontis, R. (2020). Keep calm and dream on. *Toronto Sun*, 19 April. https://torontosun.com/life/food/0419-lifenational

Destination Analytics (2020). Update on coronavirus' impact on American travel – week of May 11. *Destination Analytics*, 11 May. https://www.destinationanalysts.com/insights-updates/

Fletcher, B. (2020). It's less risky to book future travel right now than you might think. *CNN Travel*, 15 April. https://www.cnn.com/travel/article/future-travel-bookings-pandemic/index.html

Government of Iceland (2020). Testing for international arrivals could start in June. *Government of Iceland*, 12 May. https://www.government.is/news/article/2020/05/12/Testing-for-international-arrivals-could-start-in-June/

Griffin, M. (2020). Months at sea with no internet, sailing ship heads back to a 'different world'. *The Sydney Morning Herald*, 1 May. https://www.smh.com.au/world/central-america/months-at-sea-with-no-internet-sailing-ship-heads-back-to-a-different-world-20200430-p54osi.html

Groendahl, B. (2020). Vienna opens 50-euro tab for every family to promote eating out. *Bloomberg*, 13 May. https://www.bloomberg.com/news/articles/2020-05-13/vienna-opens-50-euro-tab-for-every-family-to-promote-eating-out

Grzadkowska, A. (2020). Travel insurance industry tested by disruption from coronavirus. *Insurance Business Magazine*, 27 March. https://www.insurancebusinessmag.com/ca/news/columns/travel-insurance-industry-tested-by-disruption-from-coronavirus-218177.aspx

Hajibaba, H., Gretzel, U., Leisch, F. & Dolnicar, S. (2015). Crisis-resistant tourists. *Annals of Tourism Research*, 53, 46-60.

Hall, S. (2020). Dorset crowns winner of its 2020 knob eating contest online. *inews.com*, 10 May. https://inews.co.uk/culture/dorsets-latest-champion-knob-nosher-online-2848514

Hosie, R. (2020). Iceland plans to reopen for tourists by June 15, with free COVID-19 tests on arrival. *Insider*, 14 May. https://www.insider.com/iceland-open-to-tourists-june-15-free-coronavirus-test-arrival-2020-5

Hudson, S. & Hudson, L.J. (2017). *Customer Service for Hospitality & Tourism.* Second Edition. Goodfellow Publishers Limited, Oxford, UK.

Influencer Marketing (2020). *Coronavirus (COVID-19) Marketing & Ad Spend Impact: Report + Statistics.* https://influencermarketinghub.com/coronavirus-marketing-ad-spend-report/

Lee, R. & Lockshin, L. (2011). Reverse country-of-origin effects of product perceptions on destination image. *Journal of Travel Research*, 51(4), 502-511.

Levin, J. & Oanh Ha, K. (2020). More than 90,000 cruise workers that have been stuck at sea for two months. *Fortune*, 12 May. https://fortune.com/2020/05/12/coronavirus-cruise-workers-stuck-at-sea/

Lo, Z. & Giles, O. (2020). 10 Hong Kong exhibitions to see in April 2020 - online or in person. *Tatler Hong Kong*, 13 April. https://hk.asiatatler.com/life/10-hong-kong-art-exhibitions-april-2020.

Lomas, N. (2020). Three travel startups tell us how they're responding to the coronavirus crisis. *Techcrunch*, 13 March. https://techcrunch.com/2020/03/13/three-travel-startups-tell-us-how-theyre-responding-to-the-coronavirus-crisis/

Longwoods International (2020). *Travel sentiment study wave 8.* Longwoods International, *5 May.* https://longwoods-intl.com/sites/default/files/2020-05/COVID-19%20Travel%20Sentiment%20Survey%20Wave%208%20Highlights.pdf

Mander, J. (2020). The new normal: How consumers are planning to adapt. *Globalwebindex*, 12 May. blog.globalwebindex.com/trends/the-new-normal-consumers-adapting/

Manning, E. (2020). Gurkhas rescue 109 British travelers stranded in remote parts of Nepal amid coronavirus outbreak. *Yahoo News*, 8 May. https://sports.yahoo.com/coronavirus-gurkhas-rescue-britons-nepal-125550129.html

Mazzei, P. & Robles, F. (2020). The costly toll of not shutting down spring break earlier. *New York Times*, 11 April. https://www.nytimes.com/2020/04/11/us/florida-spring-break-coronavirus.html

McKinsey (2020). A global view of how consumer behavior is changing amid COVID-19. *McKinsey*, April. www.mckinsey.com/business-functions/marketing-and-sales/our-insights/a-global-view-of-how-consumer-behavior-is-changing-amid-covid-19

Mzezewe, T. (2020). Americans stranded abroad: 'I feel completely abandoned.' *New York Times*, 18 March. https://www.nytimes.com/2020/03/18/travel/coronavirus-americans-stranded.html?action=click&module=RelatedLinks&pgtype=Article

Norris, B. (2020). UK travel insurers' coronavirus losses at least £275, says ABI. *Commercial Risk*, 24 March. https://www.commercialriskonline.com/uk-travel-insurers-coronavirus-losses-at-least-275m-says-abi/

O'Hara, C. (2020). Major Canadian insurance companies won't cover coronavirus treatment for travelers to 'high-risk' countries. *The Globe & Mail*, 14 March. https://www.theglobeandmail.com/business/article-major-canadian-insurance-companies-wont-cover-coronavirus-treatment/

Reid, C. (2020). Looking past this crisis - the future state of travel. *Hospitality Net*, 27 March. https://www.hospitalitynet.org/opinion/4097824.html

Samuel, P. (2020). How Vietnam contained COVID-19 and why its economy will rebound. *Vietnam Briefing*, 5 May. https://www.vietnam-briefing.com/news/how-vietnam-sucessfully-contained-covid-19.html/

Stoddart, A. (2020). Virtual Grand Prix Series launched to replace postponed races. *Race Tech Magazine*, 20 March. https://www.racetechmag.com/2020/03/virtual-grand-prix-series-launched-to-replace-postponed-races/

The Brandon Agency (2020). How travel is being affected by the COVID-19 outbreak. 27 March. https://www.thebrandonagency.com/blog/how-travel-is-being-affected-by-the-covid-19-outbreak/?mc_cid=4e3a14303a&mc_eid=848d5fd2d7

UNWTO (2020). COVID-19 related travel restrictions. A global review for tourism. *UNWTO*, 16 April. https://webunwto.s3.eu-west-1.amazonaws.com/s3fs-public/2020-04/TravelRestrictions_0.pdf

Vuković, K. (2020). Dubrovnik: The medieval city designed around quarantine. *BBC Travel*, 22 April. http://www.bbc.com/travel/story/20200421-dubrovnik-the-medieval-city-designed-around-quarantine

Warzel, C. (2020). They went off the grid. They came back to the coronavirus. *New York Times*, 17 March. https://www.nytimes.com/2020/03/17/opinion/coronavirus-news.html

5 The economic, social and environmental impacts of COVID-19

Figure 5.1: Aruba, 'one happy island' (courtesy of the Aruban Tourism Authority)

The Caribbean countries were among the most exposed in the world to the sudden pause in global tourism. Often dubbed the "most tourism-dependent region in the world", the Caribbean attracted more than 31 million visitors in 2019 and, for some islands, the tourism sector accounts to a colossal two-thirds of gross domestic product (GDP). "This pandemic shock is unlike any shock that these sovereigns have seen in their history," said

Julia Smith, an analyst at S&P Global Ratings. S&P expected that tourism in the Caribbean would probably decline by 60-70% from April to December 2020 compared with the previous year. As Barbara Ann Bernard from Wincrest Capital Ltd said: "Tourism money is very important for one reason: it pumps cash (dollars) into the economy. Without tourists to pay (with cash) for para-sailing, scuba diving, deep-sea fishing, taxis, groceries, etc. tourism-dependent countries risk running out of US dollars, which they need for the importation of food, fuel and for servicing debt obligations. No tourism, no cash."

Aruba, with its tagline of 'one happy island', is one of the Caribbean's most popular destinations, welcoming about 1.1 million overnight visitors and over 800,000 cruise passengers in 2019. Between them, these tourists spent $1.876 billion, accounting for 73.4% of Aruba's GDP, and generating 84.3% of all employment. So when the Aruban border was closed indefinitely to incoming visitors on March 21, 2020, the island, and its population of 113,000, lost their main source of income. Not surprisingly, by mid-April Prime Minister Evelyn Wever said that due to the COVID-19 pandemic Aruba was in a severe economic crisis.

But the impact went beyond economics. Jim Hepple, an Assistant Professor at the University of Aruba, was monitoring the situation carefully as it unfolded. "It has been estimated that unemployment in Aruba could reach at least 50% of the workforce with many people out of work and who will be without their normal income for many weeks, if not months. A major concern has to be that crime could begin to rise substantially as people attempt to obtain income to support themselves and their families." Hepple suggested that there might be other negative social consequences arising from the crisis. "It is possible that many of our best and brightest may emigrate from Aruba in search of work, leaving us with diminished skill levels in our workforce," he explained.

According to Hepple, as countries around the world eased travel restrictions in May and June, the Government of Aruba faced a huge dilemma. "The economy of Aruba is almost completely dependent upon welcoming visitors from abroad. Aruba needs visitor spending to generate income and provide employment. So, there will be enormous pressure to re-open its borders to visitors. However, once Aruba opens its borders it is inevitable that infected people from overseas will enter the country and could infect members of the local population who will in turn infect other residents. If the Government chooses to open its borders, as it must if it is to revive its economy, then it is going to have to accept that infections will continue to occur as will deaths resulting from severe cases of infection. The Government will then have to decide what is an acceptable level of mortality for its population."

However, stakeholders in the industry were keen to seen tourists return, and in May started to put measures in place for when borders were reopened. The Aruba Tourism Authority (ATA) in collaboration with stakeholders and the Department of Public Health introduced a new *Aruba Health & Happiness Code* to be implemented on the island. The new certification program was designed to ensure Aruba's visitors that the island would uphold the highest health and safety protocols throughout their stay. The objective of the program was to improve the hygiene protocols across the island and to ensure all local tourism related businesses met the required standards to earn the official 'Aruba Health & Happiness Code' seal. The Minister of Tourism, Dangui Oduber said: "As we prepare to reopen our borders, it is critical to evolve and innovate as a tourism destination. We want all visitors to feel reassured in traveling to our 'one happy island', knowing we worked together as a nation to implement the highest health and safety protocols through every step of their journey."

Jim Hepple said that because the US was by far the primary source of visitors to Aruba, the recovery of Aruba's tourism industry would be heavily dependent upon recovery in demand from that country. "It might be possible to diversify our source markets, but this will take aggressive marketing and having enough airlift in place and could take a considerable amount of time," he said. In May, the ATA did begin more aggressive marketing, introducing, for example, a 'Happily Ever After Guarantee' to provide peace of mind for those wishing to book a wedding or honeymoon in 2021. Couples could book their special occasions with participating hotels and resorts with the reassurance that they had the option to postpone should there be any coronavirus-related issues. "With nearly 316 million people under lockdown in the US alone, the travel and wedding industry has been massively impacted," CEO of the ATA Ronella Tjin Asjoe-Croes said. "We curated our 'Happily Ever After Guarantee' to offer peace of mind during these times of uncertainty for couples to start planning the wedding or honeymoon experience of their dreams on our 'one happy island'," she said.

Aruba has seen dramatic dips in tourism revenue in the past and recovered: after 9/11; again after the infamous disappearance of Natalee Holloway; and then as an after-effect of the world economic crisis in 2008. So the industry is resilient, and tourism on the island will return at some point, albeit in a different form. The current projection for the average occupancy for Aruba's hotels is a very gradual trajectory of recovery that culminates in 55% occupancy by December of 2020. It may be many years before Aruba becomes one happy island again.

Sources: Bernard (2020); Hepple (2020); Janczewski (2020); Rumball (2020)

Introduction

As mentioned in Chapter 1, the tourism and hospitality sector was particularly vulnerable to the COVID-19 crisis, especially in destinations like Aruba that were so dependent on tourism. This vulnerability has already been highlighted throughout the book, but Chapter 5 will look in more detail at the economic, social and environmental impacts of COVID-19. Most of the studies to date about the consequences of the pandemic have emphasized the economic impacts, so a synopsis of those studies will be provided. However, there have been significant social and environmental impacts from the crisis that have affected the travel sector, so these are also discussed in this chapter.

Economic impacts

The full effects of the pandemic on tourism economies around the world are still unknown, but the World Travel and Tourism Council predicted that the impact of COVID-19 on travel would be five times worse than the 2008 global financial crisis, with over 100 million jobs in the industry lost in 2020 (see Figure 5.2). Some countries will feel the impact harder than others. Of the world's 20 biggest economies, Thailand and the Philippines rely on tourism for more than a fifth of their GDP – 22% and 21% respectively. Two of the worst-hit countries in the coronavirus outbreak, Spain (14.9% of GDP) and Italy (13%), also depend heavily on the tourism sector (Quinn, 2020). However, the most severe economic devastation will likely be seen in the small island nations that have staked their entire economies on overseas travelers visiting their beaches and resorts. Of the top 20 countries most dependent on travel and tourism as a source of GDP, 15 are small island nations. One of these is Aruba, and as the opening case indicated, nearly 75% of the island's GDP comes from tourism.

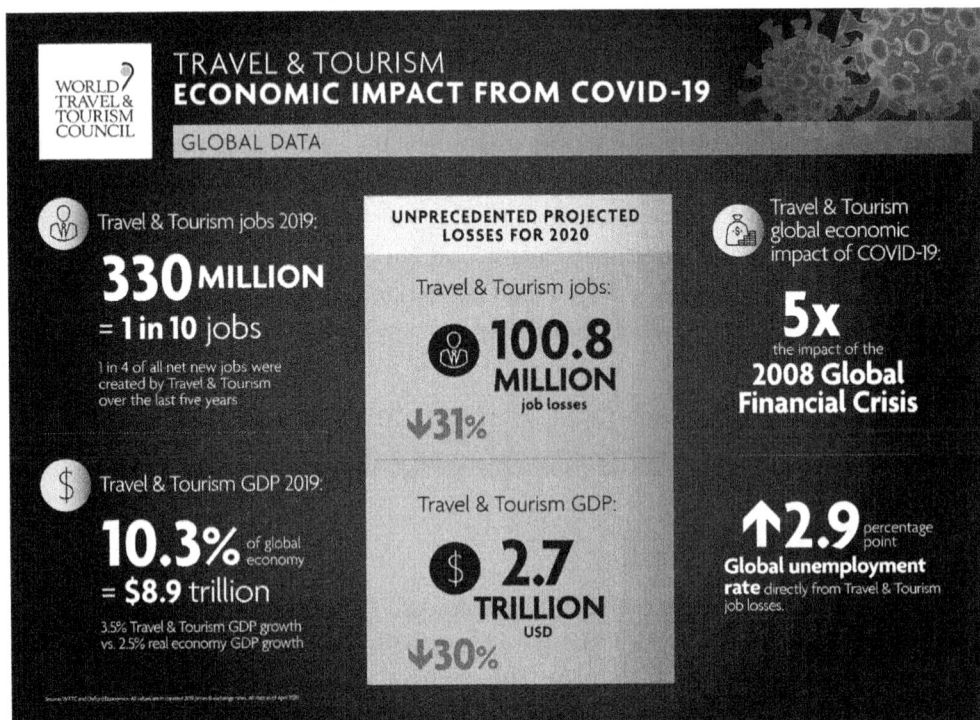

Figure 5.2: Economic impact of COVID-19 (World Travel and Tourism Council, 2020)

In absolute numbers though, tourism in the US will see the greatest losses from COVID-19, in large part because of the size of its tourism economy. International tourism receipts in the US were about $264 billion in 2019, far higher than in Spain ($81 billion) or in France ($72 billion) which ranked second and third in terms of international tourism receipts. Including the loss of domestic tourism, the US was forecast to suffer a $519 billion decline in direct travel expenditures due to the pandemic, translating into a $1.2 trillion loss in economic output – a financial impact nine times worse than 9/11 (McCarthy, 2020a). Figure 5.3 shows the sectors forecasted to experience the highest financial losses within the wider US travel industry. Food services was set to have the highest fall in direct spending at $128 billion; lodging was set to lose $112 billion while the air transportation sector was forecast to lose $97 billion in direct spending.

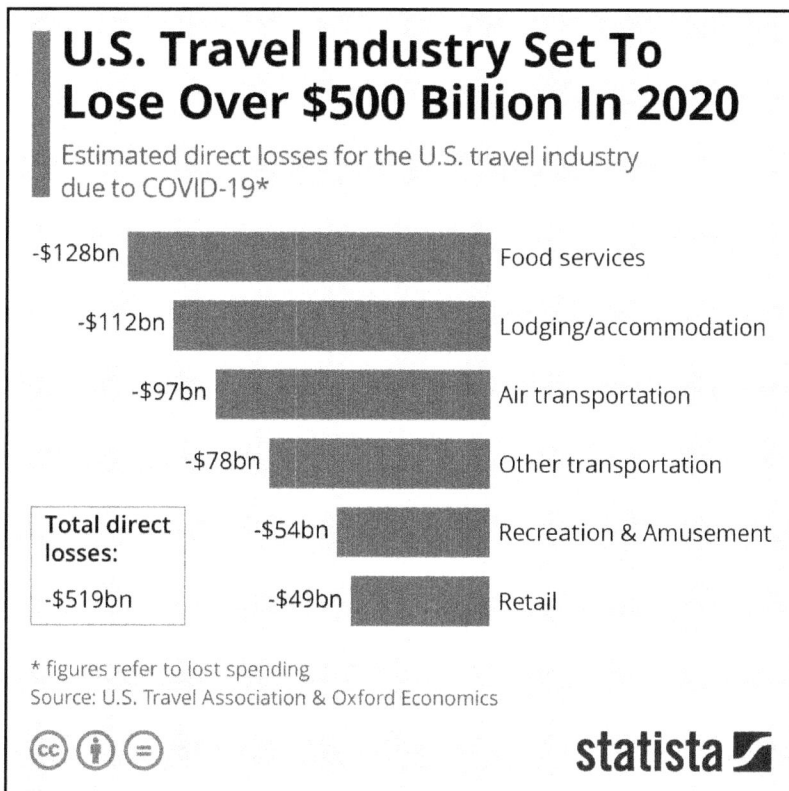

U.S. Travel Industry Set To Lose Over $500 Billion In 2020

Estimated direct losses for the U.S. travel industry due to COVID-19*

-$128bn	Food services
-$112bn	Lodging/accommodation
-$97bn	Air transportation
-$78bn	Other transportation
-$54bn	Recreation & Amusement
-$49bn	Retail

Total direct losses:

-$519bn

* figures refer to lost spending
Source: U.S. Travel Association & Oxford Economics

statista

Figure 5.3: The sectors forecast to experience the highest financial losses in the US travel industry (courtesy of Statistica)

After President Trump suspended travel from Europe for 30 days on March 11 in order to slow down the spread of the coronavirus, stock markets around the world crashed. The steep sell-off, the worst in the US since 1987, was led by the travel industry as airlines, cruise operators and online booking agencies suffered much deeper losses than the overall market. The three largest cruise operators lost more than 30% in one day, with airlines close behind (see Figure 5.4). Walt Disney

Co's value also dropped about 30% during the crisis with the company suffering a $1.4bn hit to profits in the first three months of the year. The firm shut its parks in Shanghai and Hong Kong in January, in Tokyo in February, and in the US and France in March. Its cruise lines also suspended operations during the pandemic. In April, Disney stopped paying nearly half of its workforce, furloughing more than 100,000 employees, many of them park and hotel workers.

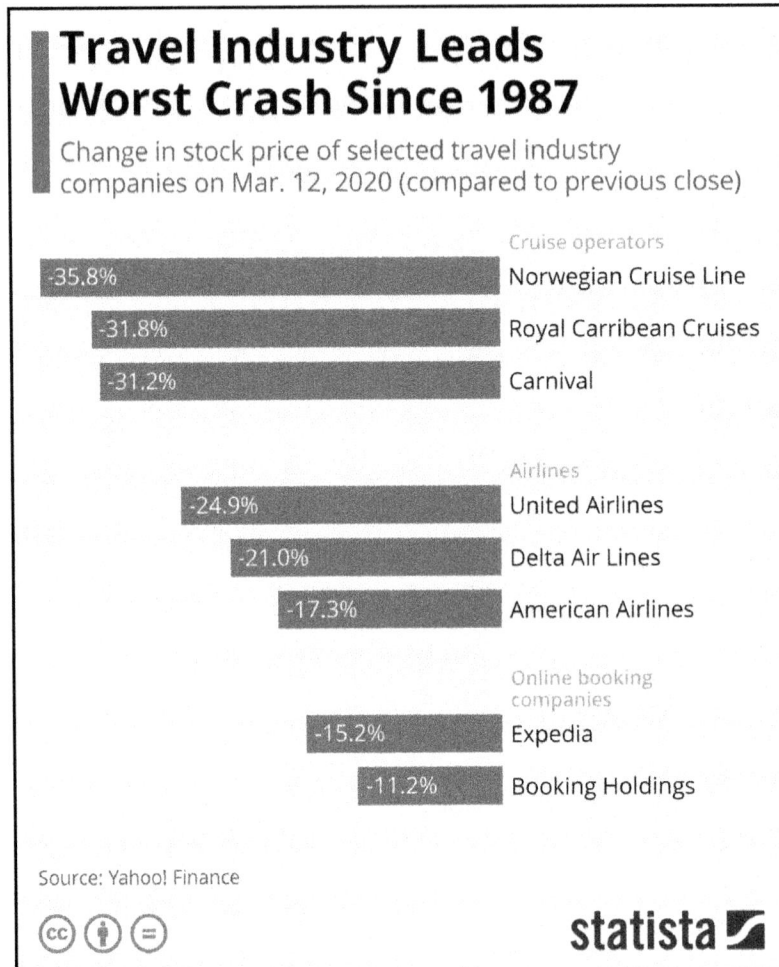

Travel Industry Leads Worst Crash Since 1987
Change in stock price of selected travel industry companies on Mar. 12, 2020 (compared to previous close)

Cruise operators
- -35.8% Norwegian Cruise Line
- -31.8% Royal Carribean Cruises
- -31.2% Carnival

Airlines
- -24.9% United Airlines
- -21.0% Delta Air Lines
- -17.3% American Airlines

Online booking companies
- -15.2% Expedia
- -11.2% Booking Holdings

Source: Yahoo! Finance

statista

Figure 5.4: Travel industry losses on the US stock market on March 12, 2020 (courtesy of Statistica)

The cancelation of mega events around the world had devastating economic impacts for host destinations. The postponement of the Tokyo Summer Olympics was referred to in Chapter 1, but another big event canceled due to COVID-19 was Expo 2020 Dubai. The six-month, multibillion-dollar global innovation fair, set to be the largest such event ever staged in the Arab world, was expected to attract some 24 million visitors from more than 200 countries starting from October 2020. Like the Olympics, the event was postponed to 2021.

The consequences of smaller events being canceled – like Coachella music and arts festival in California – were just as significant. Such events directly support jobs and generate revenue for local businesses by pulling in tens of thousands of attendees – providing a substantial economic boost to the neighboring communities. In western Canada, the Calgary Stampede was canceled for first time in almost a century, due to pandemic. The event normally draws more than one million visitors each year, and according to the Conference Board of Canada, it pumps $540 million into the provincial economy annually. On average over the past five years, the summer event has brought in $79.2 million in gross revenue and turned a profit of $21.4 million after expenses (Gibson & Dryden, 2020).

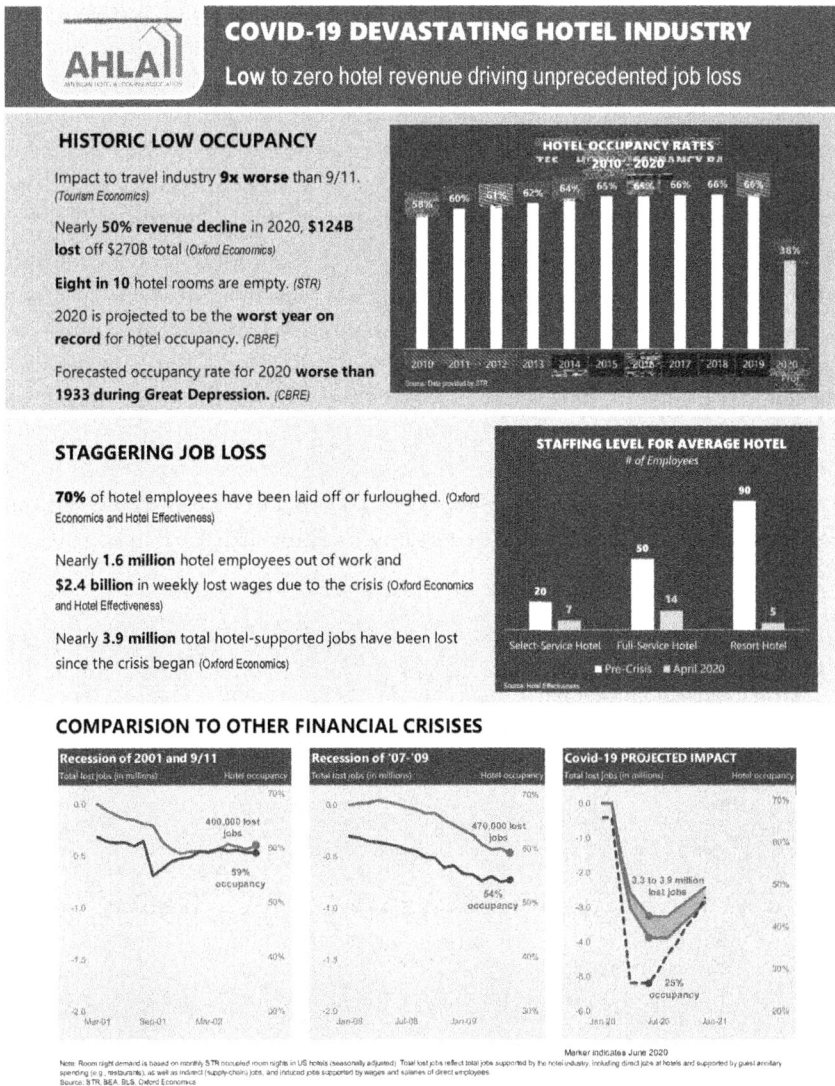

Figure 5.5: Impact of COVID-19 on the US hotel industry (AHLA, 2020)

The American Hotel & Lodging Association (AHLA) presented some data in May to show how the pandemic was harming the hotel sector in the US (see Figure 5.5). As of May 6, nearly 7 out of 10 hotel rooms were empty across the country, and hotels had already lost more than $21 billion in room revenue. The human toll was equally devastating with major hotel managers reporting significant layoffs and furloughs; nearly 3.9 million total jobs had been eliminated, with 70% of direct hotel employees laid off or furloughed. AHLA said with such low occupancy rates, many hotels might simply close their doors, putting 33,000 small business at immediate risk.

It wasn't just in the US where small businesses in the tourism and hospitality sector were reeling from a lack of visitors. In Australia, the industry had become reliant on a growing Chinese market, so they felt the pinch early on in the pandemic. The Chief Executive of Destination Gold Coast, Annaliese Battista, said China was the Gold Coast's largest international market, injecting about half a billion dollars into the economy each year, with a peak around lunar new year. Battista said it would be the small mum-and-dad businesses that would suffer the most. "Of the 3,500 tourism operators we have on the Gold Coast, the vast majority are those small- to medium-size operators," she added (Taylor, 2020).

Likewise, in Mexico City, the impact of COVID-19 was particularly painful for small operators in the tourism sector. Over 90% of Mexico's businesses are 'microenterprises,' small-scale operations with workers or owners completely dependent on each day's earnings. Informal operations, like *mariachis* (street musicians) or cozy mezcal bars, employ over 70% of the working population and account for about 50% of the country's GDP. The vulnerability of this grassroots economy was one argument for the country's belated response to the pandemic. Mexico, unlike wealthier nations, has millions of citizens living without a social safety net in an already fragile economy (Najum, 2020).

Figure 5.6: Mexico City, where the impact of COVID-19 was particularly painful for small operators in the tourism sector (photo by Daniel Lerman on Unsplash)

In the UK, the BBC reported on the impact of the pandemic on Stratford-upon-Avon businesses during the crisis. The birthplace of William Shakespeare attracts up to six million visitors annually, but the town's reliance on tourism meant it was going to be "badly hit by a perfect storm", said Councilor Daren Pemberton, deputy leader of Stratford-upon-Avon District Council. Almost overnight in March, the majority of Stratford's businesses closed. When Prime Minister Boris Johnson announced on March 16 that people "should avoid pubs, clubs, theatres and other such social venues", the Royal Shakespeare Company closed the Swan Theatre, and many other attractions and venues followed suit. The Shakespeare Birthplace Trust cares for William Shakespeare's family homes, and the trust expected to bring in $12 million in 2020. It said its income was now predicted to fall by $9-10 million. "The impact of COVID-19 on Stratford-upon-Avon has been immediate and plain to see," said its CEO, Tim Cooke. "It is home to the world's richest Shakespearian heritage, which the trust cares for, a jewel in the crown of 'Brand Britain' and the economy of the West Midlands, attracting six million visitors a year. Now it lies empty" (BBC News 2020a).

Figure 5.7: Stratford-upon-Avon - "badly hit by a perfect storm" (photo by Bobbie M on Unsplash)

Reopening after the March-May pandemic lockdown was not easy for many small businesses. In Stratford, Nigel Lambert, who runs restaurants Vintner, Lambs and The Opposition, said he had concerns about how they would operate if distancing measures continued once the doors were open. "I don't think restaurants and bars will be able to operate until there is a vaccine," he said. "If we reopen with social distancing, we still have to pay rent, lighting, power, wages and yet we will probably be doing 20% of the turnover we would normally do. It is not viable, it wouldn't work" (BBC News, 2020a). Across the pond in Lansing, West Virginia, Roger Wilson, Chief Executive of Adventures on the Gorge, which offers whitewater rafting trips, was pondering whether or not he could operate tours in the summer of 2020 with far fewer people than normal to allow for social distancing. Even if authorities allowed it and he could make it work financially, he felt customers would ultimately cast the deciding vote. "If no one comes, well, then the answer's pretty simple," he said (Scott & Pleven, 2020).

Businesses were also facing additional costs of starting up again, often having to comply with new government regulations on health and safety. "There is the expense of reopening to few customers, safety concerns for our staff, the reality of half capacity seating, and stringent cleaning and sanitizing practices," said Kathy Johnson, owner of Coyotes restaurant in Banff, Canada (Ellis, 2020). Banff was hard hit by the COVID-19 pandemic, with 85% of the local workforce having to be laid off. The iconic Banff Springs Hotel closed its doors for the first time since the Second World War.

Figure 5.8: The iconic Banff Springs Hotel which closed its doors for the first time since the Second World War (photo by Kieran Taylor on Unsplash)

Some governments or local councils provided support to help businesses cope with the demands of starting up again. In Portugal, for example, to help businesses cope with the sanitary demands of reopening, the government provided €5000 per company through its 'Adaptar' program to cover the purchase of disinfecting materials, masks, signs, and acrylic screens. The program was introduced to offer support for micro, small and medium-sized companies in adjusting to new regulations.

Another challenge for many businesses wanting to start up again as lockdown restrictions eased was staffing. In Banff, for example, many of the 5-6,000 people laid off during the COVID-19 crisis were temporary foreign workers who returned to their countries of origin. The mass exodus of foreign workers was a concern for businesses that relied heavily on them to operate in the peak spring and summer months. Daryn McCutcheon, who owns tour company Banff Adventures, said staffing would be a major concern if he and his team were able to get back to business before the end of the season. "We've got roughly a 120-day season and there will come a point where even if they said we could go in August, that the operating startup costs are so high you have to draw a line in the sand somewhere and decide, can we actually operate at all?" Despite the fact that some of his tours can run in the fall and winter months, up to 70% of McCutcheon's annual revenue is earned in the summer (Condon, 2020).

Not all employees in the tourism and hospitality sector wanted to go back to work when businesses started up again. In the US, one labor analyst said during the pandemic that a typical full-time American worker would earn $2,300 more through four months of unemployment than he would while working (Doescher, 2020). "The unemployment benefits are so generous that in many places workers are telling their bosses they'd rather be unemployed than return to their jobs," said Sean Kennedy, executive vice president of public affairs at the National Restaurant Association. "It's not that these workers are lazy, they're just making the best economic decision for their families" (Morath, 2020).

Unfortunately, some tourism and hospitality businesses were forced to close doors for good due to the pandemic. By May 2020, several airlines around the world had collapsed (Stotnick, 2020), Hertz rental car company had filed for bankruptcy, a large percentage of hoteliers in tourism destinations like Greece said they were facing bankruptcy (Wood, 2020), and OpenTable's CEO, Steve Hafner was predicting that one in four restaurants in the US would not be able to reopen (Pesce, 2020). In the UK, restaurants were hit hard after closing in March as part of Britain's coronavirus lockdown and, in May, the owner of restaurant chains Café Rouge and Bella Italia appointed administrators. The chains' owner, Casual Dining Group – whose brands also include the Las Iguanas chain – employed about 6,000 people. The UK's casual dining chains had a tough few years even before the coronavirus pandemic arrived. Some well-known names, including Jamie Oliver's restaurant empire, the burger chain Byron, and the Chiquito and Frankie & Benny's owner had either closed venues or had to put in place emergency financial measures.

Social impacts

With such devastating economic impacts from the COVID-19 crisis, it was inevitable that there would be consequences for societies dependent on tourism for their livelihoods. In Spain, it was anticipated that the human cost in bankruptcies and unemployment from COVID-19 would be 'astronomical' (Martin, 2020). In poor rural areas in many emerging countries, the impact was most acutely felt, as tourism often constitutes the only alternative to declining farming opportunities (Hudson & Hudson, 2017). The president of the World Bank, David Malpass, was warning in May that millions of people would be pushed into "extreme poverty" by the crisis. He expected global economic growth to shrink by 5% in 2020, with poorer countries feeling the brunt. "Millions of livelihoods have been destroyed and healthcare systems are under strain worldwide," he said. "Our estimate is that up to 60 million people will be pushed into extreme poverty – that erases all the progress made in poverty alleviation in the past three years" (BBC Business, 2020). Even during the first few months of the pandemic, some countries were too poor to go into lockdown. The high court in Malawi, for example, struck down the country's 21-day lockdown, not because they didn't think they needed it to stop the spread, but because too many people were too poor to be prevented from going out to make enough money to eat and pay for electricity, water, and basic housing (Poon Tip, 2020).

The COVID-19 crisis also took its toll on mental health. According to a United Nations report in May 2020, a mental illness crisis was looming after millions of people worldwide had been surrounded by death and disease and forced into isolation, poverty and anxiety by the pandemic (see Figure 5.9). In the US, from mid-February to mid-March 2020, prescriptions for antidepressants and anti-anxiety medications rose by about 19% and 34%, respectively, according to a report from pharmaceutical company Express Scripts. In the UK, levels of anxiety more than doubled during lockdown, with people's most common concerns related to their well-being, their work, and their finances (BBC News, 2020b). "The isolation, the fear, the uncertainty, the economic turmoil– they all cause or could cause psychological distress," said Devora Kestel, director of the World Health Organization's mental health department (Kelland, 2020). The UN report highlighted several regions and sections of societies as vulnerable to mental distress – including children and young people isolated from friends and school, and healthcare workers who were seeing thousands of patients infected with and dying from the new coronavirus. Others were impacted by misleading and biased media coverage and inconsistent leadership messaging about COVID-19. The mental health of many Chinese travelers and students, for example, was negatively influenced due to the outbreak having been labelled "Chinese virus pandemonium" (Zheng, Goh & Wen, 2020).

Figure 5.9: United Nations recommendations on things to do to avoid the blues during the pandemic (courtesy of Unsplash).

As travel restrictions were eased after lockdown, there was evidence of resident anxiety in destinations that started welcoming tourists once again. Such anxiety or irritation towards tourists is not new, and has been well-documented in the tourism literature (Doxey, 1976; Wall & Mathieson, 2006). But the pandemic created a dilemma for destinations as they sought to kickstart their tourism economies, and at the same time had to reassure their residents that it was safe to invite visitors into their communities once more. After Boris Johnson announced an easing of the UK's lockdown rules in May, residents at many beauty spots across the country, including Cornwall and Snowdonia, made it clear that they did not want visitors while the coronavirus was still a risk. Many even put up home-made signs urging non-locals to go home.

There were also tensions between residents and tourists in Hawaii. Locals felts that tourists were showing a disregard for their home. "People will always see this place as their playground. And in this moment, as a Native Hawaiian, this is very reflective of many historical circumstances, where people from outside of the islands have come in and caused real harm to the native population. It's not always with direct intent to do so, but the impacts, especially on Hawaiian people, are very real," said Troy Kane, a Waimānalo neighborhood board member and community representative (Pachelli, 2020). Even into June 2020, the state's tourism website was dissuading visitors from coming, saying "as a small remote island community, our residents are particularly vulnerable to the COVID-19 crisis. Hawaii Governor David Ige has asked that you postpone your trips to Hawaii to give us the opportunity to address this health crisis."

Despite taking a toll on mental health, there were positive social outcomes from the pandemic. The lockdown period led to many acts of kindness around the world, from delivering soup to the elderly in the UK, to an exercise class held for quarantined residents on their balconies in Spain, free concerts on YouTube, food donations, free handmade PPE for essential workers made by swiftly established sewing groups around the world. In Canada, the word "caremongering" was used to describe such efforts to help vulnerable populations where, facilitated via social media, the altruism was arranged online and the hashtags provided a permanent record of all the good happening in different communities across the country – an uplifting read in anxious times (Gerken, 2020).

Two tourism entrepreneurs in Spain formalized such altruism by launching a campaign to assist both the travel sector and essential workers at the same time. Ian Rutter and Andrew Watson, owners of B&B Casa Higueras in Andalucia Spain, created #MyTravelPledge, an online platform that offered free vacations to healthcare workers once the pandemic had subsided. Accommodation providers around the world were invited to join the movement and give back to frontline workers. "Seeing how frontline health workers are having to fight this virus, it gave us the inspiration to give them something to look forward to when the virus does eventually subside," said Rutter (Jadah, 2020). Accommodations as far away as Boothbay Harbor in Maine, New England participated in the initiative where

Figure 5.10: Ian Rutter and Andrew Watson, creators of #MyTravelPledge (courtesy of Ian Rutter)

Mark Osborn, owner of Topside Inn, said: "We have been looking for ways to give back to all of the people who are giving so much to keep us all safe, often putting their own health at risk. In an effort to begin to do this, we have teamed up with a growing number of international lodging properties to provide much needed breaks to these healthcare workers and their families once travel restrictions have been lifted" (Boothbay Register, 2020).

The pandemic was responsible for a number of societal changes. Dictionaries were updated as certain words and phrases became commonplace to reflect life during the pandemic, such as 'social distancing', 'self-quarantine', 'contact tracing', 'self-isolating', and of course COVID-19. The Oxford Dictionary's executive editor Bernadette Paton said that it was "a rare experience for lexicographers to observe an exponential rise in usage of a single word in a very short period of time, and for that word to come overwhelmingly to dominate global discourse, even to the exclusion of most other topics" (Flood, 2020). Ironically, it wasn't long ago (in 2018) that the Oxford English Dictionary made 'overtourism' one of its words of the year – will 'undertourism' now come into vogue? Overtourism is defined as an excessive number of visitors heading to famous locations, damaging the environment and having a detrimental impact on resident's lives. A distant memory for many destinations during the pandemic. Undertourism could mean too few tourists coming to sustain the infrastructure required for successful tourism.

The 'Low Touch Economy' was the term coined by the Board of Innovation (2020) to describe the new state of society and economy permanently altered by COVID-19, a society characterized by *low-touch* interactions, health and safety measures, new human behaviors, and permanent industry shifts. Expected to last

A series of pandemic-control health measures lead to behavior shifts and economic disruption, creating a fairly unpredictable system. The longer the health measures sustain, the more fundamental the behavior shifts and economic disruptions will be.

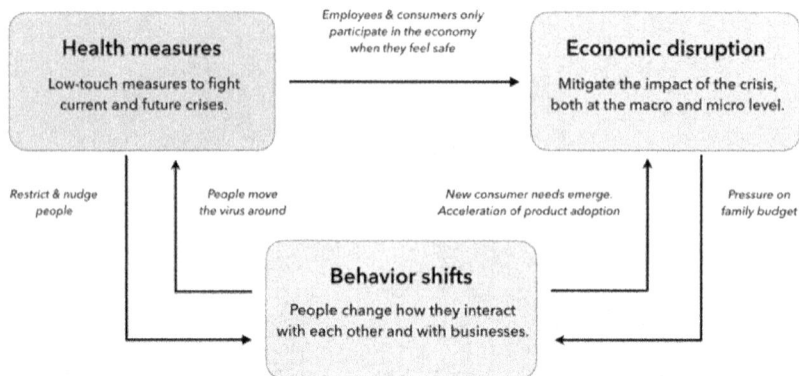

Figure 5.11: The feedback loop of the low-touch economy (courtesy of Board of Innovation)

for a few years, the Board of Innovation suggests that successful companies in this era would be those that adapted their business models to work with the different health measures and other challenges that COVID-19 presented. Figure 5.11 shows the feedback loop of the low-touch economy, whereby pandemic-control health measures have led to behavior shifts and economic disruption, creating an unpredictable system. The longer the health measures are sustained, the more fundamental the behavior shifts and economic disruptions will be.

Some of these 'enforced' low-touch interactions - like contactless payment - were easier to get used to than others. Interacting with other people in a way that was safe but didn't offend was more of a challenge. "The recent pandemic has changed everything we know, particularly about how we remain socially connected," said Bhavna Jani-Negandhi, a clinical psychologist in private practice. "People have risen to the challenge and have tried to maintain social connections in creative ways, but at the same time it has been different and it can be hard to adjust to the new ways of being social" (Stokel-Walker, 2020). Eyebrows were raised when France's health minister advised citizens to stop kissing due to COVID-19. The traditional *bise*, kiss, is used in France and some others parts of Europe as an everyday greeting by friends, family members, colleagues and dignitaries. "Neither a kiss nor a handshake: how to greet each other at the time of the coronavirus?" questioned the Novel Obs, a prominent Paris magazine. But this was one move that was not 'unprecedented'. In England in the 15th century, King Henry VI banned kissing in an effort to stymie the spread of the bubonic plague.

One key social trend that emerged from the crisis was the massive, swift and efficient transition most office workers made to continue doing their jobs from home. Some believe that this will be one important change that society keeps after the pandemic as workers (and employers) realize the benefits (McCarthy, 2020b). Companies like Facebook said they were planning to shift towards a remote workforce as a long-term trend, and New Zealand's PM Jacinda Arden suggested introducing a four-day week saying flexible working would be the new normal.

However, working from home did not suit everyone. Patrick Basset, Chief Operating Officer AccorHotels Upper Southeast & Northeast Asia, said he had seen more people renting hotel rooms in order to work *away* from home. "We have one hotel in Tokyo, the Pullman Tamachi, that has seen huge success with people coming to work in the hotel and renting space in the hotel." Basset was looking at how the Accor group could change hotels to adapt to other new needs. "Perhaps we don't need to restart with full kitchens in our hotels – we might want to outsource food-delivery to guests who want to eat in their rooms. We need to be relevant – for the next six months the main markets will be domestic and they want different services than international visitors" (QUO, 2020).

Amid the COVID-19 pandemic, people also transitioned to socializing 'virtually' from home, using platforms like TikTok and Zoom. The video-sharing social network, TikTok, exploded on to the scene during the pandemic, reaching a world full of people craving fun and interaction while at home. It became a global social media phenomenon (1.5 billion monthly users by May 2020) with everyone from doctors and nurses to celebrities sharing short clips of themselves lip-syncing to music. Virtual dating also increased during lockdown with platforms like Tinder, eHarmony, OKCupid and Match reporting a big rise in video dates (Shaw, 2020).

But perhaps the company that saw the most dramatic growth in use was video conferencing provider Zoom. Figure 5.12 shows how much Zoom's valuation shot up during this unusual period in history. As of May 15, 2020, Zoom's market capitalization had skyrocketed to $48.8 billion, more than the seven biggest airlines put together! The airline industry has seen revenue fall in total value by 62% since the end of January (Ghosh, 2020).

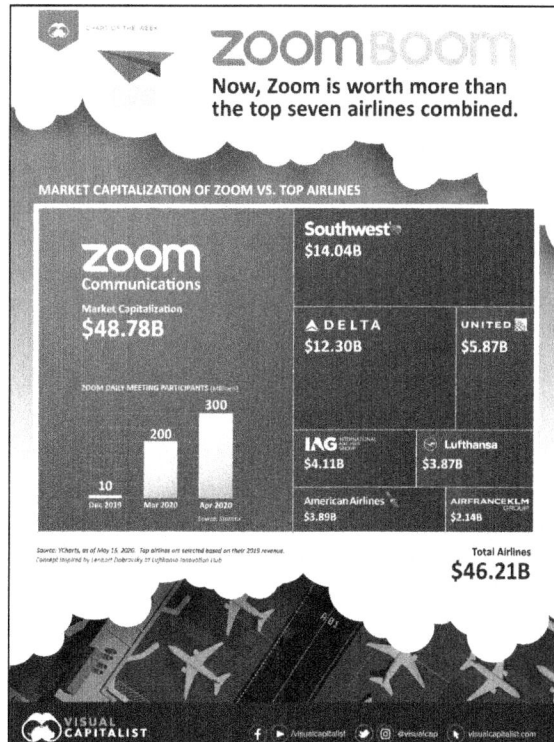

Figure 5.12: Market capitalization of Zoom versus the top airlines (courtesy of Visual Communications)

Las Vegas, recognizing that more people were relying on video conferencing to stay connected, created a series of virtual backgrounds for Zoom (see Figure 5.13). Featuring iconic Las Vegas locations, the customized virtual backdrops allowed Zoom users to infuse some fun into their everyday video chats. Co-workers could move their nightly happy hour from their kitchens to front and center at the Fountains of Bellagio; family reunions could go from the couch to overlooking the Las Vegas Strip from the High Roller; and friends could still get the party started by spending their Saturday nights on the virtual dancefloor of a Vegas nightclub.

Figure 5.13: Las Vegas Zoom backgrounds (courtesy of LVCVA)

Buying behavior also adapted to suit new needs, with online shopping moving into overdrive. At one point, Amazon announced it could not keep up with consumer demand, and delayed the delivery of non-essential items (Jones, 2020). Figure 5.14 shows e-commerce purchases in March 2020 versus those a year before. Many businesses reliant on a buoyant travel sector saw a dramatic fall in sales. Luggage sales, for example, fell 77%, cameras 64% and women's swimwear 59%. On the other hand, sales of toilet paper increased 190%, largely due to panic buying – a common human response to crisis.

Panic buying is not caused by shortages necessarily, but more by fear. At its root is a fear of scarcity, and this fear is self-fulfilling, because the more people anxiously stockpile, the more others get infected by the panic and the faster the product runs out. According to Steven Taylor, a clinical psychologist and author of *The Psychology of Pandemics*, there are parallels between the ways people have been behaving in 2020 and the way they behaved during earlier pandemics such as the Spanish flu of 1918, when there was panic buying of Vicks VapoRub. The difference now, Taylor observes, is that the panic can escalate much faster via social media and online news (Wilson, 2020).

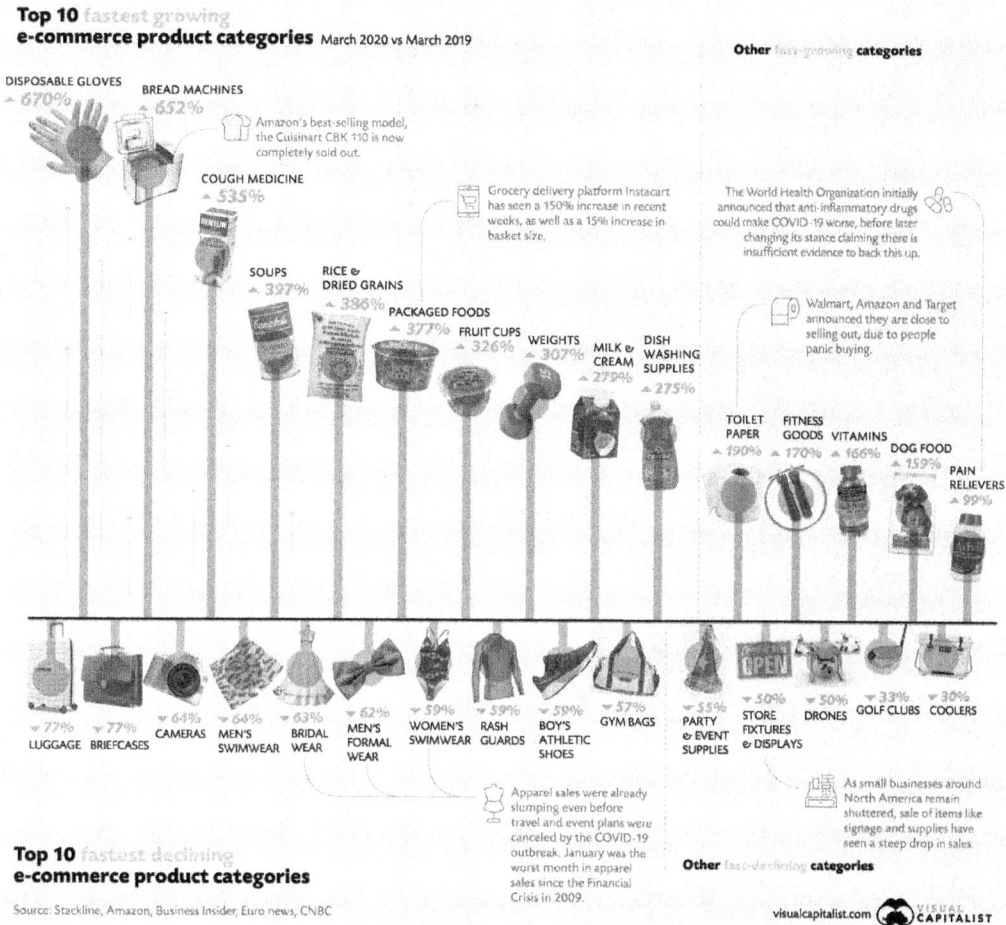

Figure 5.14: e-commerce purchases March 2020 vs March 2019 (courtesy of Visual Capitalist)

Environmental impacts

As mentioned in Chapter 3, some countries like Australia and New Zealand were promoting the fact that COVID-19 was giving the world a well-needed environmental rest and so would be even better when it was time to travel again. The media showed images of fish swimming in the unusually bright, clear and empty waters of Venice; of coral reefs in the Great Barrier Reef getting a much needed respite; and of bears in Yosemite National Park whose numbers had reportedly quadrupled during the pandemic. Certainly, the coronavirus lockdowns had had an effect on daily carbon emissions, causing a 17% drop globally during peak confinement measures by early April 2020 – levels last seen in 2006.

However, scientists said the brief pollution break would likely be 'a drop in the ocean' when it came to influencing climate change. Professor Corinne Le Quéré of the University of East Anglia in the UK who led the analysis said: "Population

Figure 5.15: COVID-19 gave coral reefs a much needed respite (photo by Fezbot2000 on Unsplash)

confinement has led to drastic changes in energy use and CO2 emissions. These extreme decreases are likely to be temporary though, as they do not reflect structural changes in the economic, transport or energy systems. The extent to which world leaders consider climate change when planning their economic responses post-COVID-19 will influence the global CO2 emissions paths for decades to come" (Rice, 2020).

Some commentators, like Jennifer Morgan, Executive General for Greenpeace, suggested that COVID-19 was an unmissable chance to put people and the planet first. She said at the time: "People in power today must step back from the way things are currently decided and instead act on the answers to these two questions in dealing with COVID-19, and the climate emergency: What do we need to do to create a more resilient and fair system that protects the most vulnerable people? How can we move the focus of our economies away from practices that put us in exposed situations like pandemics and the climate emergency, and rather put the emphasis on the well-being of people and the planet, understanding how interconnected they are?" (Morgan, 2020). The Financial Times published an editorial saying: "Given the scale of the economic damage wrought and the prospect of mass unemployment, policymakers face a difficult balancing act: do they preserve the status quo and rely on fossil fuels to revive their stricken economies or launch new policies to promote a green economic recovery" (Financial Times, 2020).

Simon Birkett, founder and director of Clean Air in London, an advocacy organization, suggested that the temporary experience of cleaner air brought about by the lockdowns could lead to a change in behavior and a demand to keep it. "People so accustomed to pollution they hardly think about it may realize 'actually, I really do quite enjoy clean air: do you think we could get it, or keep it?'" he said (Gardiner, 2020). However, to get healthier air for the longer term means shifting to clean energy and transportation, and economic troubles often prompt governments to loosen health-protective and environmentally-protective regulations rather than tighten them. But there were signs of a shift in policy in some countries. One promising environmental outcome of the crisis in Europe was that Air France promised to halve its domestic emissions of carbon dioxide within four years, in return for state loan guarantees to survive the pandemic.

Bruce Poon Tip, founder of G Adventures, published a short instabook during the pandemic called *Unlearn*. In the book, Poon Tip points to mounting criticism of the travel industry pre-crisis over issues like overtourism and flight shaming, and tourists who consume culture rather than contribute to it. Like those quoted above, Poon Tip says "this travel pause we're taking now is our chance to hit Reset". He urges travelers to be hyper-conscientious in a post-pandemic world. "I want to challenge everyone who travels to 'unlearn' what they think they know. We have the opportunity to use this reset to be more conscious and to think about how we can improve, both as individuals and as a wider travel community. We all have the ability to create positive change and to transform more lives in the future," Poon Tip advises (2020). He argues that in the future the emphasis should not be on sustainable tourism that focuses exclusively on the health of a destination, but rather community tourism as the benefits ripple far beyond the destinations that travelers visit. Community tourism includes recognition of a company's employees, suppliers, agency partners, small business owners, micro-enterprises, customers, social followers, and travelers.

Others were hoping that the sustainable or ethical travel movement would survive the COVID-19 crisis. "Sustainable travel has been on the rise for years now," said Shannon McMahon, editor at online travel magazine SmarterTravel.com. "Not only is a global pandemic unlikely to change that – it could even make traveling sustainably seem more important than ever." Kelley Louise, the founder and executive director of Impact Travel Alliance, also sees the sustainable travel movement keeping pace. While mass tourism is linked to climate change, overtourism, and conventional travel experiences, she said, sustainable tourism offers a healthier alternative for communities and for the planet: "My hope for the industry is that, when the pandemic subsides, we'll be able to explore the world with a renewed sense of mindfulness, curiosity, and appreciation" (Gibbens, 2020).

Only time will tell if this is the case, but many European cities did use the crisis to re-evaluate their transportation policies. The city core of Milan for example was partly remodeled to give over 22 miles of road space previously used by cars to bikes and pedestrians. Cars that were allowed into the center had to adhere to a new reduced 20 mph speed limit. The aim was to make traffic more fluid and give pedestrians more space to spread out safely. Recognizing the correlation between high air pollution and high mortality rates for COVID-19, cities like Milan were reluctant to promote the use of cars that do so much damage to air quality. Brussels went a stage further. From May 4, the Belgian capital's entire city core was made a priority zone for cyclists and pedestrians, one in which cars could not exceed a speed of 12 mph and had to give way in the roads to people on foot or on bikes.

Figure 5.16: Many parts of the world witnessed a cycling boom during the pandemic (photo by Victor He on Unsplash)

Like Europe, North America also witnessed a cycling boom during the pandemic. For many, bicycles became a symbol of freedom during lockdown – an opportunity for mental as well as physical release from the confines of life at home. The National Association of City Transport Officials (NACTO) said they saw an "explosion in cycling" in many American cities. Eco-Counter, which collects bike data, said that bicycle counts "significantly increased" across most of North America compared to usual. Ken McLeod, policy director at The League of American Bicyclists, suggested the trend could continue as more people commuted to work by bike after the pandemic. "Hopefully we'll see cities and governments embracing that, and making sure that it can be done safely by providing infrastructure" (Bryant, 2020).

Cities across the world also started redesigning outdoor environments to allow for additional space for socially-distanced dining. As lockdown restrictions were eased, most restaurants and cafés had to operate at 50% capacity when they reopened, and often needed more outdoor seating to make them viable. In New York, noted architect David Rockwell created a template for outdoor dining that

Figure 5.17: Melba's in Harlem. One of the restaurants Rockwell was working with to envision various outdoor seating options (courtesy of the Rockwell Group)

he made public for establishments to use if they received permits. Rockwell – who has designed everything from Broadway shows to KAOS Nightclub in Las Vegas – is especially well known for his work in restaurants such as New York's Avra Madison and Catch Steak. "We've been exploring adaptable and portable designs that extend the inner dining space to sidewalks and beyond," said Rockwell. "We've been inspired by work across the country and globe. Mostly, we've tried to utilize designs and materials that can be adapted to reflect the diversity of streetscapes in the City" (Krader, 2020).

For the travel sector, there were of course some negative environmental consequences of the COVID-19 crisis. While coyotes, bobcats and bears might have been reclaiming Yosemite National Park in the US, animals elsewhere in the world were in danger. The collapse of tourism in Southeast Asia meant that wildlife in Cambodia and Thailand was under threat. The temporary closure of Cardamom Tented Camp in Cambodia, for example, meant that forest patrols by rangers in Botum Sakor National Park were likely to be suspended. The rangers' equipment, food and wages were provided in entirety by the Golden Triangle Asian Elephant Foundation (GTAEF) and Cardamom Tented Camp, which both depended on tourism for their income – and consequently had none due to the COVID-19 shutdown. "We are extremely proud of the rangers' efforts over the last six years. They have all but stamped out bush meat poaching, the illegal wildlife trade and land grabbing on the concession," said John Roberts, Director of Elephants and Conservation at GTAEF. "However, the rangers' excellent work would almost immediately be reversed should patrolling stop in the next few weeks" (ScottAsia Communications, 2020).

Elsewhere, conservationists in Thailand were saying that more than 1,000 elephants faced starvation in Thailand because the crisis had slashed revenue from tourism. The absence of visitors meant that many caretakers were struggling to afford food for Thailand's 4,000 captive elephants. Lek Chailert, founder of the Save Elephant Foundation, said "if there is no support forthcoming to keep them safe, these elephants, some of whom are pregnant, will either starve to death or may be put on to the streets to beg" (Hatton, 2020).

Wildlife was also under threat in many parts of Africa during the pandemic. "The COVID-19 pandemic is putting conservation under enormous pressure," said Luke Bailes, founder and executive chairman of Singita, a collection of luxury reserves and lodges across the continent. "Africa's wildlife is gravely at risk if eco-tourism stops funding conservation work. If tourism collapses, the ripple effect could threaten to wipe out decades of proactive conservation work on the continent" (Holland, 2020). Dr. Jennifer Lalley, conservation director and co-founder of Natural Selection, a portfolio of owner-operated safari lodges, was just as pessimistic. "Remove tourism, and I shudder to think of the habitat destruction and decimation of wildlife populations that would ensue alongside extreme poverty. It's the very reason we got into this game." Natural Selections lodges in Botswana, Namibia and South Africa were all closed during the pandemic.

Case study: Italy: "It was like having the rug pulled out."

For many years, Italy has been one of the most popular tourist destinations in the world. In 2019, the country welcomed over 216 million tourists – a very high number for its 60 million population. That year, international tourism spending in Italy totaled $46.6 billion (versus $38.7 billion in 2014) – about 13% of GDP – and showed no signs of slowing down. A country with diverse offerings including 55 UNESCO World heritage sites, fabulous beaches and high quality ski resorts, Italy depends on its visitors with approximately 4.2 million people employed in the tourism sector.

But as the COVID-19 lockdowns triggered massive unemployment and economic hardship for millions of people across the world, Italy was one of the worst-affected nations. The country was the first to declare a nationwide lockdown, and the tourism and hospitality sector felt the impact immediately. Venice, which was nearing recovery in the Carnival season following a tourist lull after record flooding in November, saw bookings drop immediately after regional officials canceled the final two days of celebrations, unprecedented in modern times. "The shock of canceling Carnival really woke everyone up," said Matteo Secchi, head of the tourist group Venessia, referring to the city's overreliance on tourism. "It was like having the rug pulled out."

For Fulvio De Bonis, CEO of Rome-based luxury tour operator Imago Artis, the impact of the lockdown was just as shocking. "In January and February, we'd basically doubled our number of visitors compared with the same period in 2019. And we had a stellar 2020 ahead of us – or so we thought. Then, within three days of the onset of the coronavirus epidemic here, our bookings had fallen by some 80%. Virtually everyone who was booked from March onwards started canceling or postponing. It was literally a nightmare come to life. We've worked so hard to build this brand, and then suddenly: 'Is this real? Can this be?'"

As a consequence of the lockdown, tourism revenues in Italy fell by 95% in March, and the country's small business federation CNA expected 25 million fewer foreign visitors through to September. The National Tourism Agency was forecasting a €20bn fall in income compared with 2019. This impact did not include the damage that other related businesses, such as restaurants or small shops, would suffer. The government provided incentive programs to lighten the burden of those who earn their lives from tourism, but many argued that these measures were not enough. President of hotel federation Federalberghi, Bernabo Bocca, called the incentives insufficient, and asked all levels of government to adopt urgent measures to guarantee cash flow to tourism operators to protect jobs and avoid "the collapse of an industry".

However, by the end of May, efforts were under way to rescue the summer tourist season. Italy's tourism minister, Dario Franceschini, said the country was working hard to strike a balance between safety concerns and the reopening of tourism facilities. "It won't be easy, but we'll see it through," he said. Popular Italian tourism destinations were looking at measures they could adopt to find this balance. The regional government of Sardinia, for example, was working on a scheme that would require tourists coming to the island to have a document showing that they have tested negative for COVID-19. The laboratory test would have to have been conducted within a week prior to the tourist's arrival. "This way we hope to relaunch our tourism sector in June," said the island's governor, Christian Solinas. "Whoever boards a plane or a ferry will have to show a health passport along with their boarding pass and their identity document. I am sure that it will work fine: we will preserve health and save our economy at the same time. Now everything has to be done to boost tourism – it is the biggest source of income for Sardinia."

Other islands in Italy, including Capri, Ischia and Panarea – all popular high-end tourist destinations – were considering similar measures. The Mayor of Ischia also suggested installing multiple floating platforms off beaches that would allow couples or families to enjoy the sun and sea but remain at a distance from other tourists. The 6ft-wide platforms would be equipped with loungers and an umbrella.

In Venice, a victim of overtourism in the last few decades, some suggested this could be a time to rethink the future of tourism in the city. Venice's deputy mayor Simone Venturini said it could be time to consider a softer model of tourism, even if it meant physically limiting the number of visitors. "This will be an opportunity to move towards intelligent tourism. With tourists who take the time to understand and get away from the frenetic

Figure 5.18: Venice, Italy, where some said the pandemic was an opportunity to rethink tourism in the city (photo by Henrique Ferreira on Unsplash)

tours of other times," he said. Every year, as many as 30 million tourists from all over the world descend on Venice, pumping up to $2.5 billion into the local economy, according to the Italian Tourism Ministry.

Jane da Mosto, who heads non-profit group We Are Here Venice, saw the pandemic as a turning point for the city, and envisioned a new Venice emerging in the post-pandemic world. "The new Venice I dream of after this is like it is now, just with more residents," she said. "The problem for Venice isn't the lack of tourists, it's the lack of permanent residents. And with more residents, the city will reflect more the Venetian culture and the wonderful lifestyle that this extraordinary city offers and future visitors to the city will be able to enjoy Venice more." The population of Venice has dropped from 175,000 after World War II to just over 52,000 today.

Asked if there were some positives for Italy that have emerged from the crisis, Fulvio De Bonis from Imago Artis said. "The unity - it's inspiring - that's the only one. You've seen the videos of Italians singing with each other from their balconies in the evening. The coronavirus is bringing us together – making us move forward in the same direction, something we haven't always done before." For Angelo Presenza, a professor of tourism at the University of Molise, this patriotism may translate into an increase in domestic tourism, which will help the industry get back on its feet. "It seems to be the only possible scenario at the moment for several reasons such as the restrictive measures imposed for traveling abroad, the sense of fear that still exists, and a renewed patriotism".

Sources: Henley (2020); Kirkman, (2020); Bongarra (2020); Armstrong, M. (2020); Nadeau (2020)

■   References

AHLA (2020). *COVID-19's Impact on the Hotel Industry*. American Hotel & Lodging Association. https://www.ahla.com/covid-19s-impact-hotel-industry

Armstrong, M. (2020). Venice considers a new tourism model after COVID-19 lockdown. *Euronews*, 21 April. https://www.euronews.com/2020/04/19/venice-considers-a-new-tourism-model-after-covid-19-lockdown

Bernard, B.A. (2020). Caribbean tourism has been decimated by COVID-19. But the private sector can cushion the blow. *World Economic Forum*, 7 May. https://www.weforum.org/agenda/2020/05/caribbean-tourism-has-been-decimated-by-covid-19-but-the-private-sector-can-cushion-the-blow/

BBC News (2020a). Coronavirus: Stratford-upon-Avon's tourism trade hit hard by lockdown. *BBC News*, 3 May. https://www.bbc.com/news/uk-england-coventry-warwickshire-52446679?intlink_from_url=https://www.bbc.com/news/business&link_location=live-reporting-story

BBC News (2020b). Coronavirus: Money worries in pandemic drive surge in anxiety. *BBC News*, 4 May. https://www.bbc.com/news/uk-52527135

BBC Business (2020). Coronavirus: World Bank warns 60m at risk of 'extreme poverty'. *BBC Business News*, 20 May. https://www.bbc.com/news/business-52733706

Board of Innovation (2020). *The Winners of the Low Touch Economy*. https://www.boardofinnovation.com/low-touch-economy/

Bongarra, F. (2020). Italy mulls 'health passport' to help tourism recover from COVID-19 pandemic. *Arab News*, 30 April. https://www.arabnews.com/node/1667576/world

Boothbay Register (2020). #MyTravelPledge campaign to benefit COVID-19 'frontliners'. *Boothbay Register*, 7 May. https://www.boothbayregister.com/article/mytravelpledge-campaign-benefit-covid-19-frontliners/133734

Bryant, M. (2020). Cycling 'explosion': Coronavirus fuels surge in US bike ridership. *The Guardian*, 13 May. https://www.theguardian.com/us-news/2020/may/13/coronavirus-cycling-bikes-american-boom

Condon, O. (2020). Economy in Banff has 'collapsed' as COVID-19 wipes out tourism industry. *Calgary Herald*, 26 April. https://calgaryherald.com/news/economy-in-banff-has-collapsed-as-covid-19-effectively-eliminates-tourism-industry/

Doescher, T. (2020). COVID-19 rocks hospitality industry as economy sheds 701,000 jobs. *The Heritage Foundation*, 3 April. https://www.heritage.org/jobs-and-labor/commentary/covid-19-rocks-hospitality-industry-economy-sheds-701000-jobs

Doxey, C.V. (1976). When enough's enough: The natives are restless in Old Niagara. *Heritage Canada*, **2**(2), 26-27.

Ellis, C. (2020). Banff restaurants gearing up for safe reopening. *Rocky Mountain Outlook*, 14 May, 6.

Financial Times (2020). The virus fight opens up a climate opportunity. *Financial Times Editorial Board*, 15 May. www.ft.com/content/eb683e52-95d0-11ea-abcd-371e24b679ed

5

Flood, A. (2020). Oxford dictionary revised to record linguistic impact of Covid-19. *The Guardian*, 15 April. https://www.theguardian.com/books/2020/apr/15/oxford-dictionary-revised-to-record-linguistic-impact-of-covid-19

Gardiner, B. (2020). Pollution made COVID-19 worse. Now, lockdowns are clearing the air. *National Geographic*, 8 April. https://www.nationalgeographic.com/science/2020/04/pollution-made-the-pandemic-worse-but-lockdowns-clean-the-sky/

Gerken, T. (2020). Coronavirus: Kind Canadians start 'caremongering' trend. *BBC News*, 16 March. https://www.bbc.com/news/world-us-canada-51915723

Ghosh, I. (2020). Zoom is now worth more than the 7 biggest airlines. *Visual Capitalist*, 15 May. https://www.visualcapitalist.com/zoom-boom-biggest-airlines/

Gibbens, S. (2020). Will the sustainable travel movement survive coronavirus? *National Geographic*, 21 April. https://www.nationalgeographic.com/travel/2020/04/will-sustainable-travel-survive-coronavirus/

Gibson, J. & Dryden, J. (2020). 'It was mandatory': Calgary Stampede cancelled for 1st time in almost a century due to pandemic. *CBC*, 23 April. https://www.cbc.ca/news/canada/calgary/calgary-stampede-covid-19-2020-announcement-1.5542680

Hatton, C. (2020). Coronavirus: Thai elephants face starvation as tourism collapses. *BBC News*, 31 March. https://www.bbc.com/news/world-asia-52110551

Hepple, J. (2020). *What Will Aruba's Tourism Industry Look Like in the Future?* Unpublished report from the University of Aruba.

Henley, J. (2020). Covid 19 throws Europe's tourism industry into chaos. *The Guardian*, 2 May. https://www.theguardian.com/world/2020/may/02/covid-19-throws-europes-tourism-industry-into-chaos

Holland, M. (2020). With safari tourism on hold, locals and animals are at risk. *Conde Nast Traveler*, 29 April. https://www.cntraveler.com/story/with-safari-tourism-on-hold-locals-and-animals-are-at-risk

Hudson, S. & Hudson, L.J. (2017). *Marketing for Tourism, Hospitality, and Events.* Sage, London.

Jadah, T. (2020). New campaign offers healthcare workers free vacations once the pandemic ends. *Daily Hive*, 17 April. https://dailyhive.com/mapped/casa-higueras-campaign-healthcare-workers

Janczewski, M. (2020). Aruba announces new "Aruba Health & Happiness Code" certification program. *Visit Aruba News*, 20 May. https://www.visitaruba.com/news/general/aruba-announces-new-aruba-health-happiness-code-certification-program/

Jones, K. (2020). The pandemic economy: What are shoppers buying online during COVID-19? *Visual Capitalist*, 8 April. https://www.visualcapitalist.com/shoppers-buying-online-ecommerce-covid-19/

Kelland, K. (2020). UN warns of global mental health crisis due to COVID-19 pandemic. *World Economic Forum*, 14 May. https://www.weforum.org/agenda/2020/05/united-nations-global-mental-health-crisis-covid19-pandemic

Kirkman, A. (2020). The coronavirus economy: How my job as chief of one of Italy's top tourism companies has changed. *Fortune*, 5 April. https://fortune.com/2020/04/05/coronavirus-italy-tourism-jobs-covid-19-business-impact/

Krader, K. (2020). Famed designer draws up plan to save restaurants through outdoor dining. *Bloomberg*, 18 May. https://www.bloomberg.com/news/articles/2020-05-18/david-rockwell-draws-up-plans-for-outdoor-dining-in-covid-19-era

Martin, G. (2020). Post-pandemic travel: Spain, the second most-visited country on earth, weighs a fraught revival of its $200-billion tourism sector by the end of 2020. *Forbes*, 20 April. https://www.forbes.com/sites/guymartin/2020/04/20/post-pandemic-travel-spain-the-second-most-visited-country-on-earth-weighs-a-fraught-revival-of-its-200-billion-tourism-sector-by-the-end-of-2020/#4e7c24463a41

McCarthy, N. (2020a). US travel industry set to lose over $500 billion in 2020. *Statistica*, 21 April. https://www.statista.com/chart/21443/direct-losses-for-the-us-travel-industry-due-to-coronavirus/

McCarthy, N. (2020b). Why Americans wanted to work remotely pre-coronavirus. *Statistica*, 30 April. www.statista.com/chart/21563/top-reasons-for-remote-working/

Morath, E. (2020). Coronavirus relief often pays workers more than work. *Wall Street Journal*, 28 April. https://www.wsj.com/articles/coronavirus-relief-often-pays-workers-more-than-work-11588066200

Morgan, J. (2020). COVID-19 is an unmissable chance to put people and the planet first. *World Economic Forum*, 24 April. https://www.weforum.org/agenda/2020/04/covid19-coronavirus-climate-jennifer-morgan-greenpeace/

Nadeau, B.L. (2020). Deserted Venice contemplates a future without tourist hordes after Covid-19. *CNN News*, 16 May. https://www.cnn.com/travel/article/venice-future-covid-19/index.html

Najum, J. (2020). A pandemic quiets mariachis and tourism in Mexico City. *National Geographic*, 15 April. https://www.nationalgeographic.com/travel/2020/04/coronavirus-mexico-city-and-mariachis-fight-to-survive-the-covid/

Pachelli, N. (2020). It's beyond frustrating: Tensions peak. *The Guardian*, 20 April. www.theguardian.com/us-news/2020/apr/20/hawaii-coronavirus-covid-19-tourists

Pesce, N.L. (2020). 1 in 4 restaurants won't reopen after the coronavirus pandemic, OpenTable CEO warns. *Market Watch*, 15 May. https://www.marketwatch.com/story/1-in-4-restaurants-wont-reopen-after-the-coronavirus-pandemic-opentable-ceo-warns-2020-05-15

Poon Tip, B. (2020). *Unlearn. The Year the Earth Stood Still.* https://unlearn.travel

Quinn, C. (2020). The tourism industry is in trouble. These countries will suffer the most. *Foreign Policy*, 1 April. https://foreignpolicy.com/2020/04/01/coronavirus-tourism-industry-worst-hit-countries-infographic/

QUO (2020). *Podcasts on the Future of Travel.* https://www.quo-global.com/podcasts/

Rice, D. (2020). Coronavirus lockdowns have caused a whopping 17% drop in global carbon emissions. *USA Today*, 19 May. www.usatoday.com/story/news/health/2020/05/19/

5

coronavirus-has-caused-whopping-17-drop-global-carbon-emissions/5219885002/

Rumball, E. (2020). Aruba allowing couples to book events with pandemic postponement policy. *The Daily Hive*, 14 May. https://dailyhive.com/mapped/aruba-pandemic-postponement-policy

Scott, C.L. & Pleven, L. (2020). Summer businesses fear coronavirus lockdown means a lost season. *Wall Street Journal*, 18 April. https://www.wsj.com/articles/summer-will-come-crowds-are-still-a-maybe-11587182405

ScottAsia Communications (2020). Ecotourists staying away put wildlife in danger in Cambodia. *Journal des Palaces*, 19 May. https://www.journaldespalaces.com/en/news-57983-ECOTOURISTS-STAYING-AWAY-PUT-WILDLIFE-IN-DANGER-IN-CAMBODIA.html

Shaw, D. (2020). Coronavirus: Tinder boss says 'dramatic' changes to dating. *BBC News*, 21 May. https://www.bbc.com/news/business-52743454

Stokel-Walker, C. (2020). We'll be less touchy-feely and far more wary, but the transition will feel strange. *BBC News*, 19 April. https://www.bbc.com/future/article/20200429-will-personal-contact-change-due-to-coronavirus

Stotnick, D. (2020). Some of the world's airlines could go bankrupt because of the COVID-19 crisis. *Business Insider,* 12 May. https://www.businessinsider.com/coronavirus-airlines-that-failed-bankrupt-covid19-pandemic-2020-3

Taylor, J. (2020). 'Completely dropped off ': Australia's tourism industry braces for coronavirus crisis. *The Guardian*, 11 February. https://www.theguardian.com/world/2020/feb/11/completely-dropped-off-australias-tourism-industry-braces-for-coronavirus-crisis

Wall, G. & Mathieson, A. (2006) *Tourism. Change, Impacts and Opportunities*. Harlow, England: Pearson Education.

Wilson, B. (2020). There is plenty of food to go round, which means there is no need for panic buying. But who said our relationship with food was fully rational? *The Guardian*, 3 April. https://www.theguardian.com/news/2020/apr/03/off-our-trolleys-what-stockpiling-in-the-coronavirus-crisis-reveals-about-us

Wood, C. (2020). The coronavirus pandemic has hit the Greek tourism sector hard with 65% of hoteliers saying they could face bankruptcy. *Business Insider*, 19 April. https://www.businessinsider.com/greek-hoteliers-think-bankruptcy-likely-amid-coronavirus-crisis-2020-4

World Travel and Tourism Council (2020). *Economic Impact Report*, 28 April. https://wttc.org/Research/Economic-Impact

Zheng, Y., Goh, E. & Wen, J. (2020). The effects of misleading media reports about COVID-19 on Chinese tourists' mental health: a perspective article. *Anatolia*, **31**(2), 337-340.

6 The future of travel

Case study: Vietnam paving the way to the future

Figure 6.1: Ninh Binh, Vietnam (courtesy of Vietnam Travel, photo by Aaron Joel Santos)

In the spring of 2020, as countries around the world debated how quickly they could reopen their economies amid the pandemic, Vietnam was ahead of the curve. A national social distancing campaign that shut down non-essential businesses ended on April 22, and life in the country returned to almost normalcy. Restaurants, bars, cinemas, barbers and other shops reopened, and sporting events and festivals were allowed to resume. Almost immediately, the Ministry of Transport started to increase domestic flights and operate trains to major destinations albeit with limited passenger capacity.

All of this was possible thanks to a number of proactive, aggressive steps that Vietnam's government took shortly after the coronavirus emerged in Wuhan, China. On January 23, the Vietnamese Ministry of Health announced the country's first two cases of COVID-19: a Chinese father and son who were visiting Ho Chi Minh City, the country's largest urban area and its economic hub. That same day, the government canceled all pending flights between Vietnam and Wuhan, and a week later, it suspended air travel to and from mainland China. The land border was also closed to travelers.

At the same time, the government created a Steering Committee on COVID-19 Prevention and Control, headed by Deputy Prime Minister Vu Duc Dam, to coordinate the national response. It disseminated information to the public in an uncharacteristically transparent way, through daily text messages sent to all mobile network subscribers, frequent articles in state media outlets, banners hung on city streets, and a dedicated COVID-19 website built by the Ministry of Health. Travel writer Katie Lockhart arrived in Vietnam in January for three months, and watched the government carefully isolate at-risk communities, contact-trace citizens and quarantine anyone coming into the country. "Like clockwork, both morning and evening, we would hear updates blasted from loudspeakers on vehicles making their way through the streets. It seemed that at any one time, everyone knew where the most recent cases were, in full detail," she said.

While some observers attributed Vietnam's success to the country's authoritarian nature and past experience with SARS, Vietnam's effective response was also enabled by the country's ongoing efforts to improve governance and central-local government policy coordination. The government's policy to provide mass quarantine largely free of charge, meant that Vietnamese citizens (of whom 90% have health insurance) did not have to worry about costs from COVID-19 tests, associated hospitalization, and centralized quarantine, thereby increasing their willingness to comply with extensive contact tracing and strict quarantine measures. Transparency efforts also mitigated skepticism towards the Party-State's COVID-19 reporting. The Ministry of Health posted all reported cases online, enabling deeper analysis by data scientists and bloggers, and gaining endorsement from public health experts.

As a result of all these measures, the total number of COVID-19 cases as of June 2020 was estimated at 312, and the country had not reported a single death. These results were remarkable considering Vietnam's population of 95 million and its close geographical and economic ties with China. Politico, a US-based news organization, said that out of 30 leading countries, Vietnam had responded best to the COVID-19 pandemic in terms of health and economic impacts. Although GDP growth will be only around 2.7% in 2020, Vietnam could – unlike many other countries – escape a recession.

Naturally, Vietnam's tourism industry was significantly affected by the pandemic, with practically no international arrivals for three months. Prior to the pandemic, the industry – employing roughly 800,000 people – had been booming. Vietnam welcomed 18 million overseas visitors in 2019, topping 2018's record, while domestic tourism was also growing substantially on the back of sustained economic growth. However, although the situation was unprecedented, tourism experts in the country expected the industry to rebound faster and stronger than competitors. Vietnam relies heavily on Chinese and

South Korean tourists, which accounted for 56% of its international arrivals in 2019. This also presented an opportunity, as China and South Korea had also largely contained the pandemic. Matt Crate, Managing Director of WeSwap, the UK's largest travel money provider, said that Vietnam could lead the way in reopening the tourism sector in Asia: "There are countries across the world that have dealt incredibly well with the infection rate of the disease and should be commended, including Vietnam. These countries can lead the charge to help the world travel in safety."

Before opening international borders, Vietnam had to stimulate domestic tourism. In mid-May, they launched the "Vietnamese people travel to Vietnam destinations" program, aimed at developing specific tourism products and tours catered to the needs of local travelers during the pandemic. Meanwhile, airlines, travel agencies, resorts and hotels were offering discounts of up to 50% to encourage internal travel while incoming flights were still banned. Although tourism accounted for around 12% of Vietnam's GDP in 2019, domestic spending only made up an estimated 40-45% of this. As such, efforts were needed not only to encourage more domestic trips, but also to entice domestic tourists into spending more when they visited local destinations – not an easy task considering the pressures on household finances. "We have had to re-invent ourselves to focus directly on the local domestic market as well as regional Asian markets," said William Haandrikman, general manager of the Sofitel Legend Metropole Hanoi. That included room deals with $100 credits for food. Ronan Le Bihan, general manager of the Mango Bay resort in Phu Quoc, also said his resort needed to adapt to local tastes. "Tourist businesses targeting foreign tourists will be in trouble for a long time," said Bihan. "We can now focus on the Vietnamese market. But not all Vietnamese are interested in what we offer."

Vietnam was also working with bordering countries to promote inter-regional travel. Headed up by executive director Jens Thraenhart, the Mekong Tourism Coordinating Office (MTCO) is an inter-governmental body that promotes the Mekong region as a single tourism destination. Thraenhart pointed out that in Vietnam, like in other countries of the Greater Mekong Subregion, tourism is an important tool for poverty alleviation. "It pains me to see small businesses struggling to survive these challenging times," he said. However, MTCO was aggressively promoting the region to bring back tourists. "We are planning various innovative programs and initiatives, including our upcoming social media campaign 'Mekong Memories' to create a content cloud of past experiences to inspire people to 'Travel Tomorrow', and the new Mekong Deals platform to feature non-refundable vouchers and offers sold by travel operators to help businesses survive this crisis during this challenging time."

Sources: Personal communication with Jens Thraenhart; CBC News (2020); Heath & Jin (2020); Lockhart (2020); Nguyen & Malesky (2020); Quy (2020); Singh Maini (2020); Tatarski (2020)

Introduction

Most experts would agree that recovery from the COVID-19 crisis will be slow (see Figure 6.2), in large part due to the impact that the crisis has had on the global travel and tourism industry (Romei, 2020). Until there is vaccine, the virus will influence nearly every sector of travel from transportation, destination and resorts, to the accommodations, attractions, events and restaurants. The first section of this chapter looks at the future for these different sectors, a future heavily influenced by technology and a heightened emphasis on health and safety. The second part of the chapter focuses on a theme that has been prevalent in this book – the need for adaptability or 'COVID-aptability'. Consumer demands and behavior will be permanently altered by the pandemic, and all stakeholders in the travel industry will need to adapt. One part of adaptability is redesigning servicescapes – a necessity for many after the lockdown, and this is the subject of the penultimate section of the chapter. The conclusion looks at lessons learned from this crisis.

Most experts agree we're in for a long, slow, bumpy recovery.

First-order Low Touch characteristics (manage health crisis)
Second-order Low Touch characteristics (manage economic crisis)

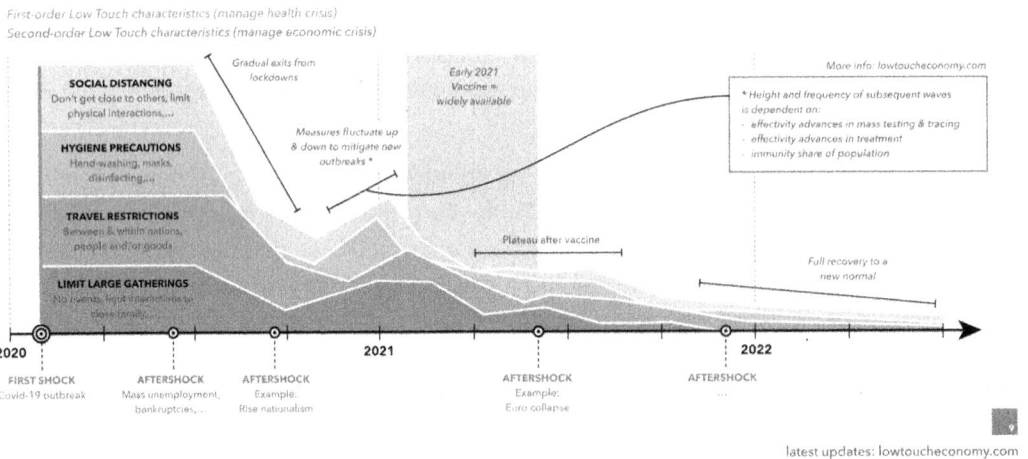

latest updates: lowtoucheconomy.com

Figure 6.2: Recovery from the COVID-19 crisis will be slow (courtesy of Board of Innovation, 2020: 9).

The future for the different sectors of the travel industry

■ Airlines

Airline strategy firm SimpliFlying has identified more than 70 different areas in the passenger journey that are expected to change or to be introduced from scratch (see Figure 6.3). These include online check-in only, contactless payments, UV sanitation of bags, hygiene-enhanced security, and health screenings or tests. Although there are no standardized decisions about making COVID-19 tests and health screenings mandatory, airports and airlines are pushing for uniform regulations. "We'll need to work with the federal government in terms of screening customers to make sure, for example, that you don't have someone getting on the airplane that has a fever," said Southwest CEO Gary Kelly. "I think that that's going to be very important" (Puckett, 2020). Emirates was the first airline to conduct on-site rapid COVID-19 tests for passengers; the quick blood test was conducted by the Dubai Health Authority (DHA) and results were available within 10 minutes. Thermal cameras, which are able to scan a crowd for a feverish temperature, were in use by May 2020 at several facilities, including Heathrow, Puerto Rico's San Juan airport, and Paine Field – a secondary airport in Seattle.

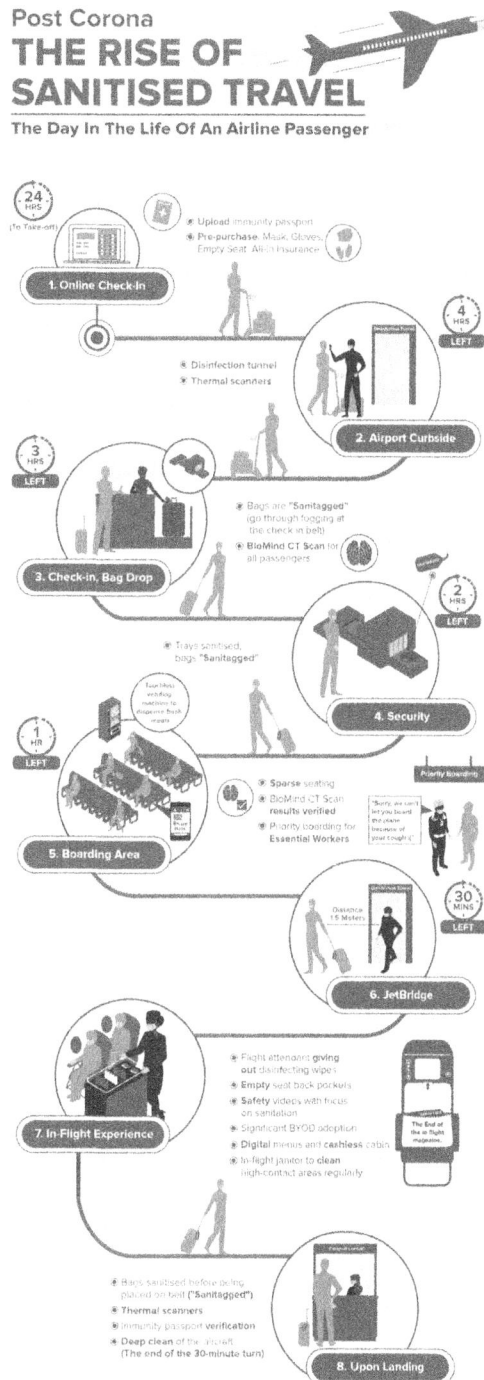

Figure 6.3: The day in the life of an airline passenger (courtesy of SimpliFlying, 2020)

6

In the future, passengers may need to present a proof of vaccine – once there is one for COVID-19 – to enter other countries. Arriving passengers will also undergo another temperature screening at their final destination and potentially even blood tests for COVID-19. Some airports, like Hong Kong and Vienna, were testing passengers for the coronavirus with a blood test before they were allowed to enter the country after the lockdown.

Airlines themselves have developed new policies and procedures to adapt to the new environment. For example, Air Canada introduced *CleanCare+*, a new program designed to reduce the risk of exposure to COVID-19 through such measures as mandatory pre-flight customer temperature checks in addition to required health questionnaires, customers care kits for hand cleaning and hygiene, and electrostatic spraying of cabin interiors. Airlines also modified inflight services for health and safety reasons. On Emirates flights, for example, magazines and other print reading material was no longer available and, while food and beverages continued to be offered on board, packaging and presentation were modified to reduce contact during meal service and to minimize risk of interaction. Carry-on items allowed in cabins were limited to laptop, handbag, briefcase or baby items. All other items had to be checked in.

It is clear that digital technologies and automation will play a critical role in the future of air travel. The need to reduce touchpoints at airports could lead to an increased use of biometric boarding that allows passengers to board planes with only their face as a passport. A number of airlines including British Airways, Qantas and EasyJet are already using the technology at their boarding gates. The shift to touchless travel and a new health safety regime will be supported by digital tools such as the Known Traveler Digital Identity initiative (KTDI). KTDI is a World Economic Forum initiative that brings together a global consortium of individuals, governments, authorities and the travel industry to enhance security in world travel. KTDI enables consortium partners to access verifiable claims of a traveler's identity data so they can assess their credibility, optimize passenger processing and reduce risk.

■ Cruise sector

As with airlines, the post-coronavirus cruise sector will also be radically different. As mentioned in Chapter 1, the cruise industry became a symbol of COVID-19, and in a recent survey almost two-thirds of over-50s said they were less likely to travel on a cruise in the future (Davies, 2020). Many experts believe that the reputational damage that has been caused by COVID-19 has been so severe that cruise companies will have to do more than offer discounts to re-attract customers. "It is essential that cruise companies address the concerns of the public and have contingency plans in place to stop a repeat of what has been seen during this outbreak. If companies do not respond sufficiently, they risk experiencing severe public backlash that could take the industry years to recover from," said GlobalData travel and tourism analyst Ben Cordwell (GlobalData, 2020a).

Genting Cruise Lines were one of the first companies to announce a new comprehensive package of health protocols for its two brands Star Cruise and Dream Cruises. The plan covered new sanitation procedures, the elimination of restaurant self-service, and new health requirements for passengers over 70. "Something like this is essential and we need to see it across the whole industry," said Cordwell. "There also needs to be full crew training to reinforce effective cleaning procedures, and to ensure the highest health and hygiene protocols are being followed." Cordwell said that much like crew, passengers will equally need to be educated on health and hygiene procedures, a duty that operators will have to ensure in the future. "Lastly, emergency procedures need to be put in place to prevent extended periods of ships being stranded at sea," he concluded. "Once they start addressing all these things, they'll be able to bring back a bit of confidence in the industry" (Future Cruise, 2020).

Analysts believe the crisis could trigger more consolidation in the cruise sector as stagnant operations and booking freezes threaten the survival of several companies. "Consolidation in the cruise industry typically comes in two ways," said Memorial University of Newfoundland professor and cruise-industry expert Ross Klein, who is also the author of the book *Cruise Ship Squeeze*. "One has been cash transfer and the other is transfer of assets." While it might still be a while before significant changes do take place, Klein expects consolidation to be one of the pandemic's most immediate consequences. "I think we're going to see the potential collapse of some companies," says Klein. "The economic impact of coronavirus is so difficult to anticipate but the worst possible scenario is one where companies are going to either go bankrupt or are going to have to mothball or sell ships" (Future Cruise, 2020). Klein also believes that the post-COVID-19 cruise sector might see "a potential redrawing of what the industry looks like and what cruise tourism looks like". For some operators, this might mean repurposing their ships and changing their market strategies – Klein even suggests turning ageing vessels into floating hotels that never leave the port.

Finally, cruise ships are likely to embrace technology at a faster pace post COVID-19. The industry has always been creative in its employment of automation and had made a head start with low-touch innovations (Hudson & Hudson, 2017a). As an example, in 2014, Royal Caribbean introduced a 'Bionic Bar' to its ship *Quantum of the Seas*; two robots who could mix drinks for passengers (see Figure 6.4). The bar was introduced more as an attraction than a labor-saving device, but it proved to be so popular that the bar has been rolled out on five other ships – *Harmony of the Seas*, *Anthem of the Seas*, *Ovation of the Seas*, *Symphony of the Seas* and *Spectrum of the Seas*. The robots can muddle, stir, shake and strain all types of drinks; the cocktail combinations are endless, with 30 spirits and 21 mixers from which to choose. The robot bartenders can produce two drinks per minute and can make up to 1,000 each day. Guests create an order on a specially programmed tablet and can keep track of their order on the digital screen next to the bar. Once the cocktail is ready, the passengers will release their drink with a simple tap of their SeaPass card or WOWband.

Figure 6.4: Royal Caribbean's Bionic Bar (courtesy of Royal Caribbean)

■ Destinations

As the case study on Aruba in Chapter 5 highlighted, until there is a vaccine, many destinations will have to find a balance between 'health and wealth' as they open their doors to tourists. As Italy (profiled in Chapter 5) opened its doors to tourists on June 1, Prime Minister Giuseppe Conte said "We're facing a calculated risk in the knowledge that the contagion curve may rise again. We have to accept it otherwise we will never be able to start up again" (BBC News, 2020). Some countries, however, were not prepared to make this trade-off, and began to create so-called 'air bridges' or 'travel bubbles' – zones of exclusive travel between two or three countries. Lithuania, Latvia, and Estonia launched the first such arrangement in May, allowing free movement for citizens of the three EU member states, at a time when travel within the larger bloc remained restricted.

The proposed travel bubble between Australia and New Zealand was the first to garner international attention and, according to New Zealand's tourism association executive Chris Roberts, would have the potential of working for travelers if rapid COVID-19 tests were available at airports (Buchholz, 2020). Figure 6.5 shows that Australia accounts for almost 40% of tourists and 25% of tourism spending in New Zealand, whereas Kiwi tourists in Australia make up 15% of visitors. In addition, around 60,000 people move permanently each year between the two countries.

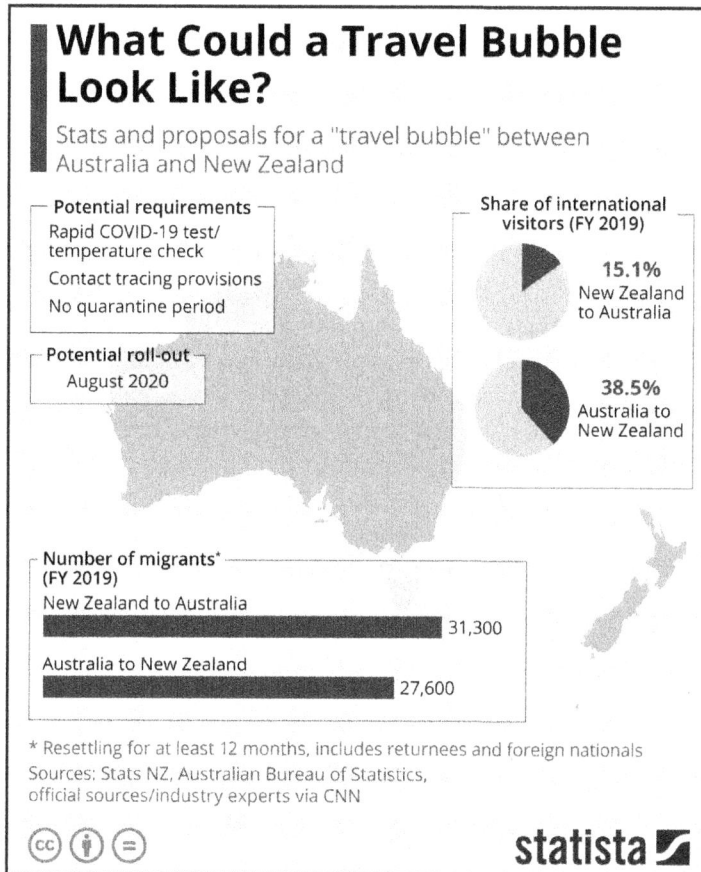

Figure 6.5: What a travel bubble could look like (courtesy of Stastica)

Travel analysts were suggesting that travel bubbles would work best between nations that have largely suppressed the coronavirus and had well-functioning institutions capable of managing future outbreaks. However, the practicalities of setting them up were complex, and within bubbles there had to be restrictions. South Korea and China opened a mini bubble in early May for business travelers, but they required visas and multiple disease tests. Other bilateral schemes could come with quarantine periods, pre-submitted itineraries, or mandatory monitoring apps. Finally, some were suggesting that rather than reunite a fractured world, the reopening of travel based on separate international circles of trust risked introducing new divisions – dividing the world into countries that had handled the pandemic well from those who had struggled (Crabtree, 2020).

Rather than create bubbles, other destinations opened up for tourists from low-risk countries only. Cyprus, for example, opened its airports on June 9 to a number of countries that were perceived to be low risk, including Germany, Greece, Israel and Malta. They still tested tourists on arrival but, in an interesting move, they pledged to cover the holiday costs of anyone who tested positive.

Officials also said a 100-bed hospital would be set aside specifically for tourists who tested positive, as well as several so-called 'quarantine hotels' for the patients' families. In mid-June, more than 10,000 German tourists were welcomed to the Spanish Balearic Islands as part of a 'safe corridor' trial. The trial came ahead of the rest of the country reopening to tourists. Through an agreement with the German tour group TUI, other operators and a number of airlines, the Germans were allowed into the Balearics as long as they stayed a minimum of five days. The holidaymakers first filled out a detailed health questionnaire before flying. On arrival they passed signs in German and Spanish reminding them to wash their hands and wear masks, then queued at a distance from each other to have their temperatures read and paperwork checked by Spanish border staff. Francina Armengol, the regional president, described the scheme as "an important step in helping to restore the Balearics's reputation as a safe, quality destination", adding that the Germans had been chosen because of their low death rate during the pandemic. Historically, Germans have made up a large proportion of Spain's tourists. Hotels were limited to 50% occupancy and had to install infra-red cameras to measure tourists' body temperatures (Jones, 2020).

Other destinations counted on collaborative marketing efforts with other countries to kickstart tourism after the pandemic. As mentioned in the opening case study, Vietnam worked with neighboring countries in the Mekong region to promote inter-regional travel. "What makes travel in the Greater Mekong Subregion unique and authentic are small local enterprises and operators," said Jens Thraenhart, executive director of the Mekong Tourism Coordinating Office (MTCO). "Over two years ago, we created the Experience Mekong Collection, which is a collection of over 300 small responsible businesses and social enterprises. This consortium, recognized as a global best practice by the UNWTO aligned to the Sustainable Development Goals, is developing into a powerful community of like-minded businesses, all looking to survive, to help each other, and being able to cross-promote fellow member businesses of the collection when travel resumes. A recent survey we did with these businesses has shown that about 84% of revenue has been lost due to COVID-19. But what is reassuring is that a large majority believe that this globally unique initiative will be able to help them through this crisis and accelerate recovery post COVID-19."

Meanwhile, destinations all over the world implemented measures to convince international travelers they were safe and could guarantee high standards of service and cleanliness. "We will be ready to offer our visitors a safe, healthy, and detoxing platform to decompress once borders begin to open," said Saeed Rashed Al Saeed, marketing director of Abu Dhabi Culture & Tourism, as the country prepared to open borders once more. Abu Dhabi was hoping that emerging destinations situated away from crowded cities would appeal to future customers. "We believe that the current global crisis will compel the international traveler to slowly test the waters across emerging destinations with less pressure, where they can enjoy healthy choices, nature, top affordable luxury hospitality, and clean facilities and attractions," said Al Saeed (Hamdi, 2020).

Some destinations created new quality standards for health and safety in order to reassure potential tourists. Chapter 3 discussed how Turismo de Portugal created a seal of approval called *Clean & Safe* to highlight tourist properties and activities that ensured compliance with hygiene and cleaning requirements for the prevention and control of COVID-19. Companies could use the *Clean & Safe* seal, either physically on their premises, or on digital platforms (see Figure 6.6). VisitBritain also introduced a 'physical distancing' standards symbol for hotels and attractions in the UK to encourage the return of visitors once lockdown measures were eased.

Figure 6.6: The Pestana Group in Portugal promoting the Clean & Safe seal (courtesy of the Pestana group)

Resorts were also having to adapt to the new environment. Club Med, a pioneer of the all-inclusive resort concept, introduced a program called *Safe Together* that was implemented in all of the brand's international resorts. Club Med worked with an International Scientific Committee comprised of a specialized team of doctors and professors, and closely monitored guest sentiment around the world. "We understand expectations on health and safety have shifted, and after a long period of social distancing travelers will want to revisit places they are familiar with and trust," said Carolyne Doyon, President and CEO of Club Med North America (Byers, 2020). The main measures in the *Safe Together* program were protective face coverings, gloves for restaurant staff, resort capacity at 65%, deep cleaning procedures, floor markings for physical distancing, temperature checks, and a doctor or nurse available 24 hours a day, seven days a week.

For destinations seeking guidance on recovery after the crisis, the UNWTO designed a *COVID-19 Tourism Recovery Technical Assistance Package* (see Figure 6.7). The package offered guidance to both public and private tourism sector stakeholders in their crisis response by, first, outlining UNWTO's range of technical assistance, and, second, detailing potential areas of intervention, including impact assessment, roles and responsibilities. The package was structured around three main pillars: economic recovery; marketing and promotion; and institutional strengthening and building resilience (UNWTO, 2020a). "These guidelines provide both governments and businesses with a comprehensive set of measures designed to help them open tourism up again in a safe, seamless and responsible manner," said UNWTO Secretary-General Zurab Pololikashvili. "They are the product of the enhanced cooperation that has characterized tourism's response to this shared challenge, building on knowledge and inputs from across the public and private sectors and from several UN agencies as part of the UN's wider response" (UNWTO, 2020b).

Figure 6.7: UNWTO's COVID-19 Tourism Recovery Technical Assistance Package

■ Accommodations

For the accommodation sector, recovery may take many years. Analysts are predicting demand for US lodging accommodations could return to pre-crisis levels in the third quarter of 2022, although a lag in average daily rate growth will stall the recovery in revenue per available room (RevPAR) until 2023 (Simon, 2020). In Asia, Smith Travel Research (STR) is suggesting occupancy may not return to 2019 levels until 2023, and RevPAR not until 2024. As mentioned in Chapter 2, to win back customers, hotels started to change operations to emphasize cleanliness and safety. The World Health Organization (WHO) issued guidance on the "Operational considerations for COVID- 19 management in the accommodation sector", and individual countries followed with their own sets of guidelines. In the US, the American Hotel and Lodging Association (AHLA) released new safety and cleaning guidelines for hotels to use when reopening amid the coronavirus pandemic, and major hotel brands like Marriott, Hilton, and Wyndham pledged to abide by the new protocols. The Malaysian Association of Hotels (MAH) began a 'Clean and Safe' campaign for hotels, and the Singapore government offered establishments the opportunity to be certified as 'SG Clean'.

Some countries went beyond guidelines, insisting on certain procedures being in place before hotels could open. In Egypt, for example, for hotels to resume operations they had to have a clinic with a resident doctor, regularly screen temperatures and install disinfection equipment. Guests had to register online, and staff had to undergo rapid coronavirus tests when entering hotels, while a hotel floor or small building needed to be assigned as a quarantine area for positive or suspected coronavirus cases. Hotels were to operate at 50% capacity, and were not permitted to host weddings or parties, organize entertainment activities, or offer open buffets.

As the hospitality sector restarted operations, hotels themselves introduced their own cleaning programs for their properties worldwide. Such programs included Wyndham's 'Count on Us', Best Western's 'We Care Clean', Hilton's 'Hilton CleanStay', Choice Hotel's 'Commitment to Clean', Marriott's 'Global Cleanliness Council' and Accor's 'ALL Stay Well' (see Figure 6.8). In order to fund new cleaning programs, many hotels – including franchise-run chains – cut amenities such as daily housekeeping visits, hot breakfast buffets and even complimentary soaps and lotions (Broughton, 2020). Technology played an important role in the roll out of new operations. Irish hotelier John Fitzpatrick, for example, was planning a series of changes for his two hotels in New York which included an electronic thermometer at the front door that took the temperature of workers, guests and other visitors to his hotels, and a robot that helped sterilize bedrooms (Hancock, 2020).

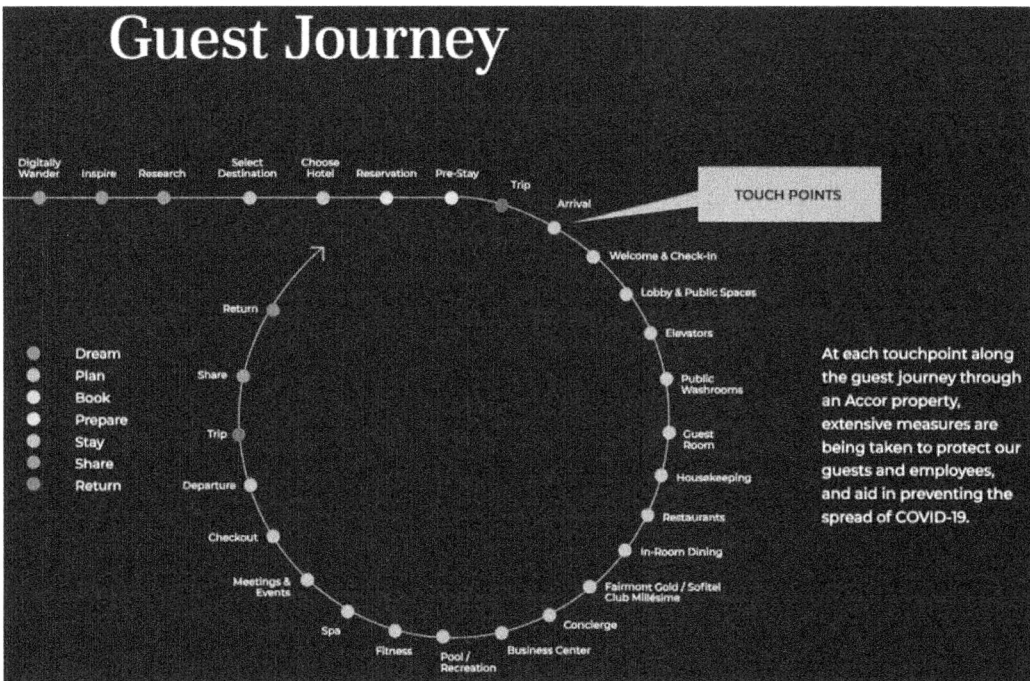

Figure 6.8: Accor's 'ALL Stay Well' program was intended to impact each touchpoint of the guest journey (Accor, 2020)

As hotels looked towards the future, it wasn't just cleanliness that was top of mind. Acqualina Resort & Residences, a luxury 98-bed beachfront hotel in Sunny Isles Beach, Florida, was planning to place an increased emphasis on customer service. "Our goal is to offer an exclusive ultra-luxury experience that will feature a heightened level of service," said Deborah Yager Fleming, CEO of the hotel (Friedman, 2020). Guests were to have dedicated 'experience managers' similar to private concierges, who would deliver personalized services. Technology usage was to be enhanced, with a new smartphone app that allowed guests to check in remotely and order food and have it delivered to specific locations.

Communication was another strategy that hotels focused on post-pandemic to keep stakeholders informed and facilitate guest and staff safety. "We'll have a lot of signage to educate employees and guests," said Bob Rauch, CEO of RAR Hospitality in San Diego, who owns five hotels in Arizona and California. He planned to provide more communication with guests before they arrive, during their stay and after their stay, as well as with employees to better meet everyone's expectations (Friedman, 2020).

To compete on cleanliness with hotels, Airbnb introduced a new enhanced cleaning initiative late in May 2020 called *The Cleaning Protocol*. The protocol was not mandatory for property hosts, but guests could filter searches for hosts who had earned the certification under the new protocol. With guidance of how to specifically clean each room, the protocol also complied with many of the Centers for Disease Control and Prevention (CDC) recommendations including a rule requiring a 24-hour vacancy period between guest reservations. "Homes have become a place of shelter, and the future of travel will also rely on a new comfort zone, with the privacy and benefits of a home away from home, without crowds or high turnovers," said Greg Greeley, president of homes at Airbnb (Rizzo, 2020). Despite this initiative, some experts were suggesting that hotels would have an advantage over Airbnb and other types of lodging as they all attempted to recover from the pandemic. "I do think hotels may have a near-term advantage," said Henry Harteveldt, a lodging industry analyst and the founder of Atmosphere Research Group, predicting that hotels would have the edge on hygiene and standardized social-distancing policies (Glusac, 2020).

■ Attractions

Casinos around the world were also preparing for a post-pandemic world during the summer of 2020. As Las Vegas opened up in June, nightclubs, buffets and large venues remained closed. Signs everywhere reminded guests of new rules: wash your hands; keep distance from others; limit your elevator ride to your sanitized room to just four people. "You're going to see a lot of social distancing," said Sean McBurney, general manager at Caesars Palace. "If there's crowding, it's every employee's responsibility to ensure there's social distancing" (Ritter, 2020). Dice were disinfected between shooters, chips cleaned periodically and card decks were changed frequently.

At some resorts guests were encouraged to use cellphones for touchless check in, as room keys and to read restaurant menus. Wynn Resorts properties and The Venetian, owned by Las Vegas Sands, used thermal imaging cameras at every entrance to intercept people with fevers. Guests received free masks at large resorts, but were not forced to use them. For blackjack dealers, bellhops, reservation clerks, security guards, housekeepers and waiters, masks were mandatory. "That's the most visual thing. Every employee will be required to wear a mask," McBurney said. At the neighboring Bellagio, where new hand-washing stations replaced banks of slot machines, casinos were limited to 50% of capacity. "You're going to see less people, by control and by design," said Bill Hornbuckle, acting chief executive and president of casino giant MGM Resorts International. His company was losing almost $10 million a day during the shutdown, he said (Ritter, 2020).

Theme parks were also having to abide by new rules as they opened their doors once more after lockdown. Shanghai Disneyland reopened in May with 'controlled capacity', temperature checks and everyone wearing masks. Lines were structured, and ride vehicles strategically loaded to promote social distancing. Shanghai Disneyland attracted roughly 80,000 people a day under normal conditions, but the Chinese government wanted capacity capped at 24,000 a day. Disney's parks elsewhere, and other large theme parks like Universal Studios, followed similar measures as they opened up (Santoro, 2020). In the first phase of the Disney World reopening, the park was able to fill to 50% capacity to ensure social distancing. Mandates largely focused on employees wearing masks at all times, getting their temperature taken before each shift, and staying home if they were feeling ill. Guests were also required to wear masks and get their temperatures taken before entering the park. Touchless hand-sanitizer stations were located all over the parks – especially at ticket booths and ride entrances – and railings and surfaces were wiped down after every use. Guidelines for large parks also included putting down tape markings for guests to stand six feet apart on each attraction line.

Technology played an important role for attractions as they reopened. When the leaning Tower of Pisa (Figure 6.9) opened its doors in Italy after a three-month closure, tourists not only had to wear face masks, but also an electronic device which sent out a signal if anyone got within a meter of anyone else. Visitor numbers were restricted to 15 at a time at the monument, compared to its usual maximum of 200. Many attractions, like Beijing Palace Museum and Kolmarden Zoo in Sweden, insisted on advance online reservations. The director of Amsterdam's world-famous Rijksmuseum, Taco Dibbits, said: "Every case is different, not just in the Netherlands, and we all have to find our own way to provide a safe experience for visitors. I think almost all museums will insist visitors buy a ticket in advance and keep to the slot they are given online" (Dowd, 2020).

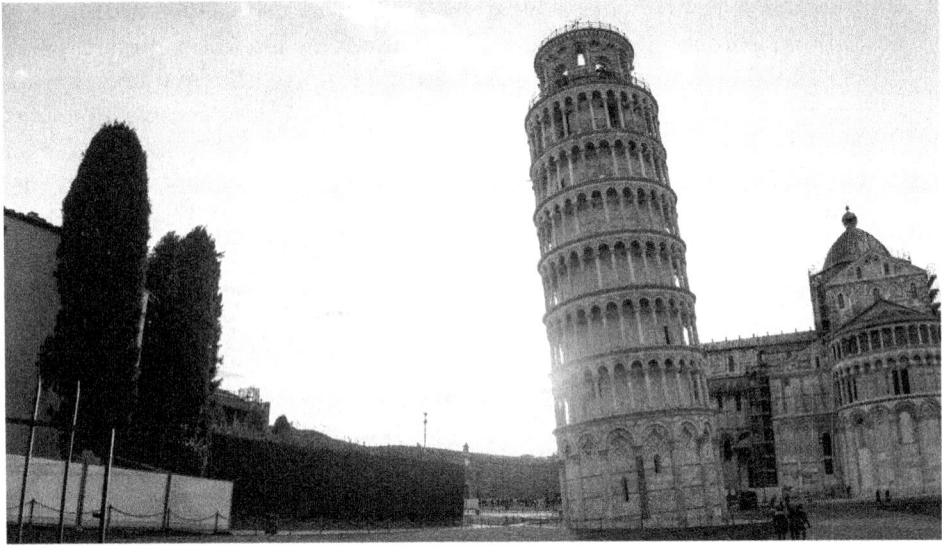

Figure 6.9: Italy's leaning Tower of Pisa (photo by Yeo Khee on Unsplash)

Some attractions – like the Birla Industrial and Technological Museum (BITM) in Kolkata, India – installed full-body disinfection stations at their entrances, and others, like the Madrid's Prado museum and the Vatican Museums, required some sort of temperature or health check upon entry. With social distancing being extremely important for any attraction, it is likely that the use of apps will become more important than ever. During the crisis, Holovis, a leading experience designer, developed a new social distancing app for attractions called Crowd Solo that acted as a virtual queuing and reservation system (see Figure 6.10).

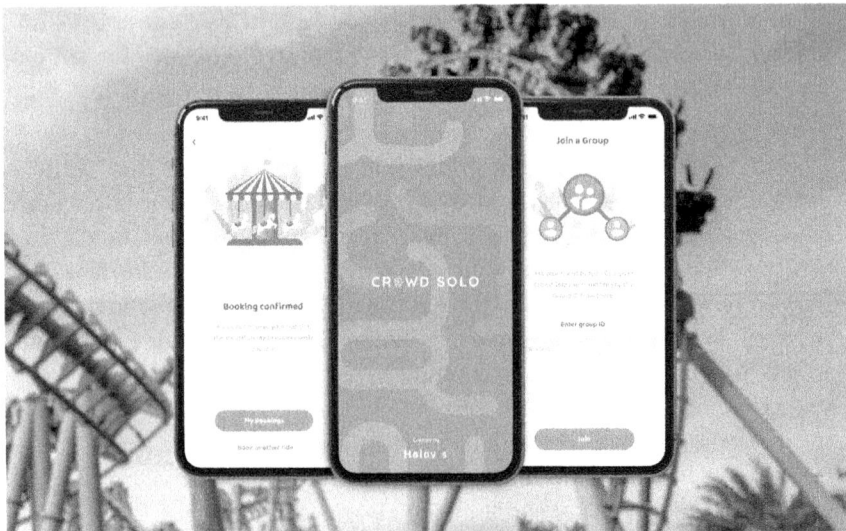

Figure 6.10: The social distancing app called Crowd Solo developed by Holovis (courtesy of Holovis)

Guests could download the app then use it as the only way to book attractions for themselves and their family throughout the day. By connecting everyone on the same platform, an exact 'state of the park' map was created. Operators could use the Operator Portal to view heat-map data in real-time that showed them how their space was performing. They could then take instant action to disperse any hotspots forming and analyze the data daily to help with expanding capacity, staffing plans and general facilities' management. The app was used for the first time in June 2020 by a theme park in Norway.

■ **Events**

As mentioned in Chapter 5, the cancelation of events around the world had devastating economic impacts for host destinations, and this is one sector that may take longer to recover than others. Jonathan Worsley, chairman of Bench Events, believes that events may change forever after the pandemic. Worsley is one of the founders and co-organizers of leading conferences for the hotel investment industry, and he had to pivot immediately during lockdown. He launched *Hospitality Tomorrow*, a virtual conference platform to support the hard-hit international hospitality community during COVID-19.

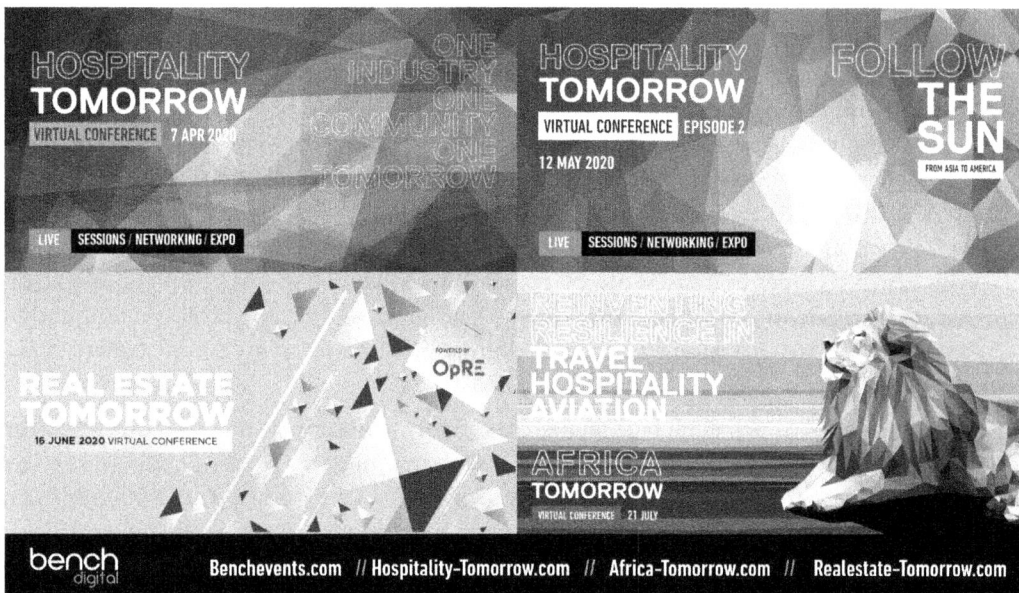

Figure 6.11: Hospitality Tomorrow (courtesy of Bench Events)

The first episode took place on April 7, 2020 and attracted 5,323 virtual attendees from 128 countries. The immersive conference experience featured industry insights by global speakers on the main 'live' stage and in break-out sessions and roundtables, as well as live face-to-face networking and a virtual Expo hosted by the event sponsors. "This was going to be a big year for us – we have had to put 10 events on hold – but as they say, crisis creates creativity," he said (QUO, 2020).

"We put a small team together to look at the future and try and resolve the needs of our clients. We looked at what other industries were doing – and found we are very traditional – other sectors are far ahead of us using technology to host events. As an outcome we created *Hospitality Tomorrow* within 17 days of the UK going into lockdown."

Worsley was not the only event organizer to migrate to online platforms. Business and professional online events increased 1,100% in April 2020 compared to April of the previous year, according to Eventbrite. "Our data is showing a significant rise in demand for online events and it's been inspiring to see the innovative ways event creators are leveraging our platform," said Crystal Valentine, chief data strategy officer at Eventbrite. "We expect online events will continue to play a big role in events post-pandemic" (Murphy, 2020). Live streamed events were already seeing an increase in engagement before the coronavirus outbreak. The video broadcasting tool Socialive got an early start by partnering with Facebook and LinkedIn when it first launched in 2016, but little did it know the company's revenue was going to skyrocket four years later. "When COVID-19 hit, we just saw an inundation of requests," said David Moricca, CEO of Socialive. "We've grown our customer base by 90% since the beginning of March and our revenue has grown 146%" (Murphy, 2020).

John Capano, senior vice president of client development at Impact XM says that for events in the future, companies and event planners may combine virtual and in-person experiences. These hybrid events would host the main event in one location, but have satellite events in other locations to avoid having thousands of people congregate in one space at a time. Jonathan Worsley agrees: "I think 70% of events that used to be face-to-face can now be done online. I just don't see financial directors of companies sending employees to events in the near future." When it comes to networking, Capano thinks that this might be one of the factors a company might have to dial down on their virtual event. "You can still do live chats, create chat rooms and virtual lounges," Capano says. "You're just not doing it as easily as before" (Murphy, 2020).

But the future of face-to-face events is not all gloomy. Research from Northstar Meetings Group in June 2020 found that many industry professionals were eager to get back to face-to-face events as soon as it was safe to do so. Although 75% of respondents intended to offer a virtual component to their live events as an alternative for those who were not willing or able to travel, only 40% of planners anticipate a high level of satisfaction with digital gatherings. "Our events are very social along with being informative," commented one respondent. "Losing the in-person experience reduces a big portion of why people attend." For many, getting up to speed on digital alternatives was a significant source of stress. "I have 30 years of live event experience," commented one survey respondent. "I'm concerned about not being able to provide the same level of expertise with virtual events" (Northstar, 2020).

As for the live entertainment industry, Live Nation CEO Michael Rapino was upbeat about the future when he was interviewed in May. "Our global diversity is our greatest strength, it always has been, and unlike sports, we have very diverse sizes of shows. We did 15,000 club and theater shows in 40 countries last year," Rapino stated. "So over the next six months, we'll be starting slow and small, focusing on the basics and testing regionally. We're going to dabble in fan-less concerts with broadcasts and reduced-capacity shows, because we can make the math work" (Shaffner, 2020). Noting that plenty of artists who can sell out arenas will be willing to play smaller-scale shows, Rapino anticipated a resurgence in theater and club shows throughout the summer in countries that were more advanced in the recovery process.

Live Nation was also looking at staging drive-in concerts which were seeing some success in Europe. In Denmark, singer-songwriter Mads Langer was one of the first to play to an audience social-distancing inside their cars at a drive-in stage erected in the coastal town of Aarhus. Interacting with the audience using a huge Zoom conference via a jumbo on-stage display, fans showed their appreciation by blowing their horns and using their windscreen wipers, while staff in hazmat suits served up movie snacks and drinks. Langer came up with the drive-in idea after staring down the barrel of having more than 40 European summer festival dates disappear in the wake of coronavirus: "I'm very much looking forward to playing normal concerts again, hug my audience and be closer to them once I'm allowed to. But in the meanwhile, I definitely think the drive-in concept is worth exploring" (Newstead, 2020).

Figure 6.12: Mads Langer playing a drive-in concert in Denmark (courtesy of Mads Langer)

■   Restaurants

Like events, the restaurant sector was decimated by the COVID-19 pandemic (Dube, Nhamo & Chikodzi, 2020), and experts were not particularly optimistic about the future. Well-known American restauranteur David Chang, who has opened more than a dozen restaurants around the world, believes the restaurant industry may never be the same after the pandemic. "My fear is the restaurants that survive are going to be the big chains, and we're going to eradicate the very eclectic mix that makes America and going out to eat so vibrant and great. I see the complete destruction of the midmarket restaurant, the mom-and-pop restaurants" (Marchese, 2020).

In an op-ed published by *The New York Times* on March 24, a group of high-profile New York chefs and restaurateurs estimated that 75% of the nation's independent restaurants wouldn't be able to open without a more robust relief package than Congress was offering. It seemed that New York's 26,000 restaurants (and their 350,000 workers) were uniquely ill-prepared to weather a prolonged shutdown. Rising rents, rising wages, rising food costs, delivery-service fees, and other factors had left independent owners with almost no cash reserves. "Over the last 10 years, restaurants have been so intensely overregulated," said Camilla Marcus, owner of West-Bourne, a cafe in Soho. "Our operating margin is 10%, at best. That's the target for a restaurant. Most businesses can't conceive of 90% going out the window. Think about that. That dynamic doesn't exist in any other industry" (Bort, 2020).

As mentioned in Chapter 5, as lockdown restrictions were eased, most restaurants and cafés had to operate at 50% capacity, but this was not feasible for some given the tight margins. "Our 34-seat restaurant can make sense economically if we're busy on all 34 seats," says Gabe Stulman, who owns several small restaurants in Manhattan. "If we're not busy on all 34 seats, it doesn't make sense. If we're not busy on 17, it definitely isn't viable. It's just not working" (Bort, 2020). In addition to capacity restrictions, restaurants have had to abide by new safety regulations. Visual cues conveying that cleanliness is a top priority are critical to establish trust with consumers going forward. In fact, diners may have witnessed a form of "cleanliness theater" post lockdown – a celebration of the actual act of cleaning the space (Barry, 2020).

Restaurants in China, which began opening up in March, gave the restaurant sector elsewhere a glimpse of its future, but not a rosy one. From fine dining restaurants to small mom and pop noodle shops, nobody in Beijing managed to make up for the overall loss of income caused by COVID-19 with delivery and takeout. And, as they opened back up, the new visible safety measures were not enough to get customer volume back to pre-virus levels (Fox-Lerner, 2020). Small obstacles to eating out – heightened risk, temperature checks, hospital-like sanitizing signs, servers with masks, less atmosphere due to big empty spaces – were proving to be a deterrent to customers (Genung, 2020).

COVID-aptability

During lockdown on April 23, English rock band the Rolling Stones released their first single in four years called "Living in a Ghost Town". The song was based on 2019 recording sessions and the band fast-tracked releasing the song due to its relevance to social distancing as a method to control the spread of COVID-19. Mick Jagger changed some of the lyrics to refer to the pandemic. The initial release was digital-only, accompanied by a music video with footage taken from across the world of empty city streets. The song was a huge success, and among its achievements on various *Billboard* charts, the song reached No. 1 on Billboard's Rock Digital Song Sales survey.

Like the Rolling Stones, travel organizations had to pivot during the pandemic and think creatively about how they adapted to the 'new normal'. As Chapter 2 showed, many organizations did pivot successfully during lockdown: restaurants focused on takeout and delivery for income; commercial airlines flew critical medical supplies and other cargo around the world; and hotels pivoted to lend a helping hand – supporting the pandemic by providing overflow capacity for hospitals, or offering people a facility to self-isolate and protect their families. However, as travel restrictions eased, organizations had to consider how they could adapt to the next stage of the pandemic – one characterized by an emphasis on health and safety, technology and a low-touch economy.

One good example of a company pivoting to cater to new demands was EmiControls, a division of TechnoAlpin, a world leader in snowmaking technology. With sluggish sales of existing products, the company pivoted to satisfy a new demand – the need for disinfection. Using similar technology to snowmaking, EmiControls provided machines – either to rent or for sale – that could distribute disinfectants both indoors and over large open surface areas. The disinfectant was atomized into a fine mist made up of tiny droplets, and this fine mist would then settle over every surface due to its slow speed of sedimentation. EmiControls also created a disinfection tunnel, a kind of open-air shower that allowed people to disinfect themselves from head to toe. In the travel sector, the tunnel has been used at attractions, airports and hotels.

Figure 6.13: The disinfection tunnel produced by EmiControls (courtesy of EmiControls)

As this example shows, the ability to adapt is a competitive advantage in business (Reeves & Deimler, 2011) and is particularly relevant during a crisis (Girneata, 2014). Global Data (2020b) suggested that because consumer demands and behavior will be permanently altered by the pandemic, all stakeholders in the travel industry will need to adapt. They say there are short, medium and long-term strategies that the industry can adopt to mitigate the impacts of COVID-19 (see Figure 6.14).

Travel & Tourism COVID-19 mitigation strategies

COVID-19 has hit the tourism industry hard. It is impossible to say when travel will return to 2019 levels, 2020 should bear the brunt of disruption
Consumer demands and behavior will be permanently altered, players at all stages of the value chain will need to adapt

Short-term strategies 6-12 months	Mid-term strategies 1-3 years	Long-term strategies 3-5 years

▪ Focus on survival.
▪ Cut costs, conserve cash, secure credit and government funding.
▪ Airlines: assess route planning and drop low-demand/low-profit routes.
▪ Lodging: focus heavily on high hygiene standards and promoting them to instil confidence in potential guests. Maintain price discipline.
▪ Cruises: focus on 2021. Encourage re-booking by offering incentives such as extra credit vouchers.
▪ Intermediaries: maintain partnerships with key transport and lodging providers.
	▪ Address oversupply in the market: M&A, footprint, and brand portfolio rationalization.
	▪ Prepare for a potential change in how customers see the world. Be prepared to adapt swiftly.
	▪ Assess how demands and expectations may change.
	▪ Adjust product and marketing strategies accordingly.
	▪ Invest in effective marketing campaigns, incorporating the learnings from the COVID-19 crisis.
	▪ Maintain the momentum of domestic tourism, gained during global travel restrictions.
		▪ Prepare for a market rebound that might require a very different product/service to that needed now within different competitive environment:
		  ▪ Possible long-term shift to reduced business travel as people have become more confident in videoconferencing.
		  ▪ Hotels should look to market hygiene standards provided by daily, professional housekeeping as this is an advantage over accommodation sharing sites.
		▪ Continue to re-evaluate and, if necessary, adapt branding and product/service. Consumer behaviour will not be static.

Figure 6.14: Travel and tourism COVID-19 mitigation strategies (courtesy of GlobalData, 2020b: 100)

The Board of Innovation (2020) in a report on the *Low Touch Economy* suggests that in the short-term, organizations may have to look for substitute products and services in order to satisfy demand. Figure 6.15 gives some examples of possible substitute services offered during the pandemic to satisfy the needs of international travelers. There are examples of these throughout the book, from the Faroe Islands' *Remote Tourism* initiative, to VisitBritain's *Bring back the love* virtual campaign, to VisitScotland's *#AWindowOnScotland* campaign targeting domestic tourism.

In an article about how businesses can move from surviving to thriving after the coronavirus, McKinsey & Co propose four strategic areas to focus on: recovering revenue, rebuilding operations, rethinking the organization, and accelerating the adoption of digital solutions. They recommended having a start-up mindset that favors action over research. They give the example of a Chinese car-rental company whose revenues fell 95% in February as the pandemic took a stranglehold on business. With the roads empty, company leaders didn't just sit back; instead they reacted like a start-up. They invested in micro–customer segmentation and social listening to guide personalization. This led them to develop a database of new potential customers. They discovered, for example, that many tech firms were telling employees not to use public transportation so they used this insight to target such firms. They also called first-time customers who had canceled orders to reassure them of the various safety steps the company had taken, such as 'no touch' car pickup. To manage the program, they pulled together three agile teams with cross-functional skills and designed a recovery dashboard to track progress. Before the crisis, the company took up to three weeks to launch a campaign; that became two to three days. Within seven weeks, the company had recovered 90% of its business, year on year – almost twice the rate of its chief competitor (Sneader & Sternfels, 2020).

Step 2: Spot substitutes, inside or outside your industry

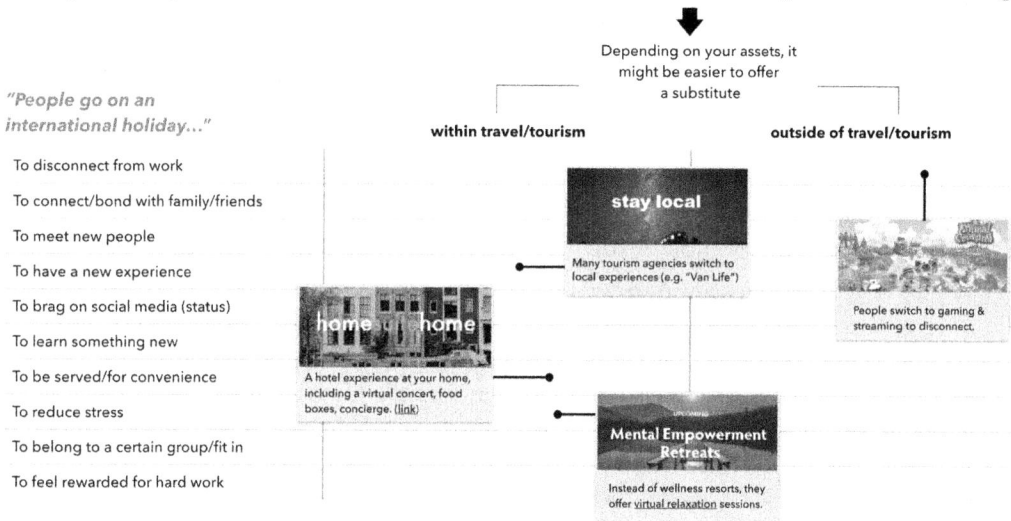

Depending on your assets, it might be easier to offer a substitute

within travel/tourism   **outside of travel/tourism**

"People go on an international holiday..."

To disconnect from work

To connect/bond with family/friends

To meet new people

To have a new experience

To brag on social media (status)

To learn something new

To be served/for convenience

To reduce stress

To belong to a certain group/fit in

To feel rewarded for hard work

stay local

Many tourism agencies switch to local experiences (e.g. "Van Life")

home sweet **home**

A hotel experience at your home, including a virtual concert, food boxes, concierge. (link)

People switch to gaming & streaming to disconnect.

Mental Empowerment Retreats

Instead of wellness resorts, they offer virtual relaxation sessions.

Figure 6.15: Substitutes for international travel during the pandemic (courtesy of Board of Innovation, 2020: 53)

The servicescape redesigned

The environment in which a service is delivered is often referred to as the 'servicescape', and is very important for tourism and hospitality products such as hotels, restaurants, and theme parks, which are dominated by experience attributes (Hudson & Hudson, 2017a). After the COVID-19 crisis, all sectors of the travel industry had to rethink – and often redesign – their servicescapes. Even municipalities had to modify outdoor public spaces. After lockdown such spaces became synonymous with freedom, providing a sense of release and respite from domestic confinement. Daniele Terzariol, deputy mayor for the town of San Donà di Piave, near Venice, said that as the city reopened, the administration "wants to use this opportunity to make our public spaces more functional and more beautiful than they were before the lockdown". To allow for socially distant gatherings, the town pedestrianized key areas of the center and launched a competition among restaurant and bar owners for the best outdoor arrangement that would enable people to come together safely – anything from movable furniture to tape art on the floor to encourage safe walking paths. "All ideas and materials are welcome, particularly when recycled and low budget," said Terzariol (Del Bello, 2020).

As destinations eased out of lockdown, visitor management of outdoor spaces became a priority for many. "What is needed is destination management to rebuild tourism more slowly and keep residents, visitors and businesses that depend on tourists happy – it's quite a balancing act" said Patricia Yates, acting CEO of Visit Britain. Some destinations put in place visitor monitoring systems to measure and regulate visitor flows in order to prevent excessive gatherings of people – especially close to popular attractions. Beaches in the Mediterranean were divided into grids to separate groups of visitors by age, as well as implementing time slots, restrictions on activities and a limited capacity (IQ Latino, 2020). Plastic or plexiglass screens were also being used by some beaches to separate bathers. On the Greek island of Santorini, owner of beach bar Demilmar Charlie Chahine built plexiglass barriers around sun loungers on the beach to protect guests against close interactions with others. "We'll do whatever it takes to make the visitor feel safe," he said (Bouras, 2020).

Servicescapes indoors were sometimes not as easy to redesign in order to ensure appropriate distancing. In Italy, for example, some people criticized solutions that were based on defensive design principles such as plastic screens to split restaurant tables, making it impossible for customers to get close to one another. Some felt that the screens were impractical and took away the joy of a shared meal (Del Bello, 2020). In Porto, Portugal, restaurateurs balked at the visual effects of expensive acrylic screens, saying they would detract from the look of their establishments and make clients feel uncomfortable. "We can't even fill the few tables we have! We're definitely not going to 'expand', or buy protective screening," one owner said (Donn, 2020). Even if the plastic screens don't take off in restaurants, the trend of open kitchens could become more prevalent as more diners express a desire to see their food being made. Showing customers that their food is handled with the highest hygiene standards is one way of restoring trust (Hudson & Hudson, 2017b).

Hotel design may also undergo significant changes post COVID-19 pandemic. Jesper Palmqvist, STR Area Director, Asia Pacific, said occupancy levels to begin with would be driven by the type of property and whether or not social distancing was possible in that hotel. "There are a lot of properties that have built-in inherent social distancing villas, and they will recover quicker than others," he said (QUO, 2020). According to hotel architect Baskaran Kolathu, all the elements and spaces in hotels in the future will be designed with minimum human contact and will drive guests to enjoy the spaces just through visual sense. "There will be more cost allocated to technology, where one can facilitate minimum human interaction," said Kolathu. He anticipates an increased use of robots in housekeeping, automated entry systems, one-card identification systems, and automatic lighting systems. The other major change that hotel design may witness is the choices of materials. "As guest facing facilities need to be designed to quickly wipe down, we may see a departure from the 'rough' aesthetics that had gained some traction over the past few years. There may be fewer carpeted areas, and the use of tiles and stones may increase," said Kolathu (Singh, 2020).

Lessons learned

■ Tourism is an important industry not to be taken for granted

Tourism's contribution to the global economy has risen over the past decade and is now responsible for about 10% of economic output, and for one in four of all new jobs created. The blow that COVID-19 dealt to the global travel and tourism industry is therefore set to do lingering damage to the world's growth, as areas that are dependent on visitors for their income struggle to reposition their local economies. "The coronavirus pandemic triggered an unprecedented crisis for the tourism economy, which has significant implications for international service trade, jobs and growth that support many local communities and regional development," said Lamia Kamal-Chaoui, director of the OECD Centre for Entrepreneurship, SMEs, Regions and Cities (Romei, 2020).

This book has highlighted how the tourism and hospitality sector was particularly vulnerable to the COVID-19 crisis, especially those destinations that were over-dependent on tourism. For these destinations, the crisis has been a wake-up call, and perhaps governments in these countries will either recognize the need for a more diversified economy, or invest in the tourism industry to make it more resilient and sustainable. The Experience Mekong Collection, referred to above, is a good example of a government initiative to build resilience in a region where tourism is an important tool for poverty alleviation.

For some tourism destinations, though, this crisis may represent an opportunity. As J.F. Kennedy was fond of saying, when written in Chinese, the word crisis is composed of two characters; one represents danger and the other represents opportunity. The empty streets of Barcelona have made local businesses and the tourist board re-evaluate their priorities. "While we couldn't continue at the speed things were, this is showing us that no tourists is no good either – there needs to be a more moderate way," said Mateo Asensioof the Barcelona tourist board. "Our first task is getting locals back out into the city, then the domestic market and our neighbors. When the international market returns, we'll focus more on specific sectors. It's an opportunity to change the rules" (Dunford, 2020).

Residents in Europe were also hoping the crisis would bring positive change to their communities. Polling data in May 2020 from 21 cities across six European countries showed a clear majority in favor of measures geared at preventing a return to pre-pandemic levels of air pollution (Posaner, Cokelaere & Hernandez-

Morales, 2020). Some, though, were missing the tourists. Wouter Vanstiphout, from Delft University of Technology in the Netherlands, said residents who live in central Amsterdam have had a rude awakening. "Now that tourism has stopped and the Airbnbs are empty, they have discovered they have no neighbors," he said. "There is no neighborhood. There is no city. If you subtract the tourists, there is nothing" (Wainwright, 2020).

Other sectors of the travel industry may see the crisis as an opportunity to change. The restaurant sector, for example, was in trouble long before the pandemic arrived, so this could be a chance to reevaluate the industry's structural inadequacies and create a more stable business model. For many, it is the pay structure that needs to change and the overreliance on cheap labor (Nunn, 2020). "One of the things this pandemic has exposed is, all of a sudden you have people making more on unemployment than they would normally," said Dave Seel, founder of Blue Fork Marketing, and cofounder and president of the Baltimore Restaurant Relief Fund (Saladino, 2020). Derek Brown, owner of Columbia Room in Washington D.C., thinks customers might need to pay slightly more to subsidize employee pay and business stability. "I don't want to raise the price," said Brown. "But I want the price to reflect the real cost of training, production and cost of ingredients" (Saladino, 2020).

■ Strong leadership is required to deal with a crisis

Martin Luther King Jr said the ultimate measure of a man is not where he stands in moments of comfort, but where he stands at times of challenge and controversy. King was almost correct. But, as Chapter 2 pointed out, it was female leadership that often made the headlines for the right reasons during the crisis. Experts say that the women's success may offer valuable lessons about what can help countries weather not just this crisis, but others in the future. Some said that female leaders during the crisis were more likely to listen to outside voices – crucial for a successful pandemic response. "The only way to avoid 'groupthink' and blind spots is to ensure representatives with diverse backgrounds and expertise are at the table when major decisions are made," said Devi Sridhar from the University of Edinburgh Medical School. Others proposed that women were more likely to follow a risk-averse strategy which proved to be successful. "What we learned with COVID is that a different kind of leader can be very beneficial," said Dr. Alice Evans, a sociologist at King's College London. "Perhaps people will learn to recognize and value risk-averse, caring and thoughtful leaders" (Taub, 2020).

As far as leadership in the travel industry is concerned, it did seem that many leaders in the industry were not prepared for this pandemic; perhaps justifiably so, as this crisis was – as it has been said so often – 'unprecedented'. However, as suggested earlier in the book, the first step to the management of any crisis is to establish a crisis management team *before* any crisis breaks out and, clearly, this was the exception rather than the rule. But what the crisis did show is that there are plenty of leaders in the travel sector who are willing to adapt and change their business models in order to survive. The end of chapter case study about restauranteur Terry Jensen is a good example of this. Such COVID-aptability will be a crucial management skill moving forward.

■ Communicating during a crisis is critical

Another lesson from this pandemic is that keeping open lines of communication during a crisis – to all stakeholders – is crucial. This book has profiled a number of organizations that understood the importance of this: Micato Safaris, Marketing Greece, Hotels.com, Auckland Tourism, the Las Vegas Convention and Visitors Authority, and more. Using a variety of communication tools – from social media to print advertising – these organizations also understood the need to change their tone of message as the crisis unfolded. One excellent example from Chapter 3 is Zermatt in Switzerland, whose tourism authority took a creative but relevant approach to keep their destination top of mind by projecting one by one the flags of different countries on to the iconic Matterhorn mountain every night during lockdown as a display of solidarity with others. The campaign generated a considerable amount of media worldwide, reaching over 700 million people.

If history repeats itself, it is likely that those organizations that remained vocal during the crisis will recover faster than others, having communicated to consumers the image of corporate stability during challenging times, and maintained or even increased 'share of mind' (Adgate, 2019; Sapient, 2020). Another tactic that proved to be successful during the crisis for tourism and hospitality providers was engaging in philanthropic activities. The final case study in Chapter 2 highlighted how hotels all over the world pivoted during the crisis to lend a helping hand, and Chapter 5 documented how individual accommodation providers joined together under the #MyTravelPledge umbrella to provide free vacations to healthcare workers.

A recent article in the *Journal of Business Research* said that such efforts of corporate social responsibility (CSR) will not go unnoticed by consumers, and will be even more important in the future. "The pandemic will teach us a lesson that 'we are all in this together', which undoubtedly will raise people's expectation of businesses being more socially responsible. Therefore, we can envision the post-pandemic period as one in which the thriving businesses are those with strong CSR commitment and effective CSR strategies and efficient implementations. Greenwash, pinkwash, and lip services will no longer survive closer consumer and public scrutiny," said the authors (He & Harris, 2020).

■ Travelers may change – forever

Just as governments may not take the travel industry for granted in the future, travelers themselves are more likely to appreciate their ability to travel. The Germans actually have a word to describe the lust for travel during lockdown – *fernweh*. Marrying the words *fern*, or distance, and *wehe*, an ache or sickness, the word can be roughly translated as 'distance sickening' or 'far woe' – a pain to see far-flung places beyond our doorstep (Farley, 2020). So consumers will be keen to travel after the crisis, although they may change their behavior. Evidence suggests that the first travelers to test the waters after lockdown were a younger generation, supporting the 'crisis-resistance tourist' theory proposed in Chapter 4. In China, for example, travelers born after 1990 accounted for more than half of the total bookings during the 2020 May Day break. This youth-led tourism recovery was expected to be replicated in other countries and regions (Liang-Pholsena, 2020).

Although some people would be desperate to travel after lockdown (He & Harris, 2020), Jens Thraenhart, executive director of the Mekong Tourism Coordinating Office believes the traveler of the future will be more cautious than before. "In the post-pandemic world, we will see a shift in preference and behavior among travelers – the public health conditions of destinations, and the hygiene standards of transportations, hotels and other tourism facilities will become a top priority; people will prefer short-haul breaks and shorter itineraries; regional tourism and economic collaborations may become driving factors for an accelerated tourism recovery. Countries have invested resources over the past decades to build and foster these collaborations. Now is the time to harvest the fruits of the hard labor."

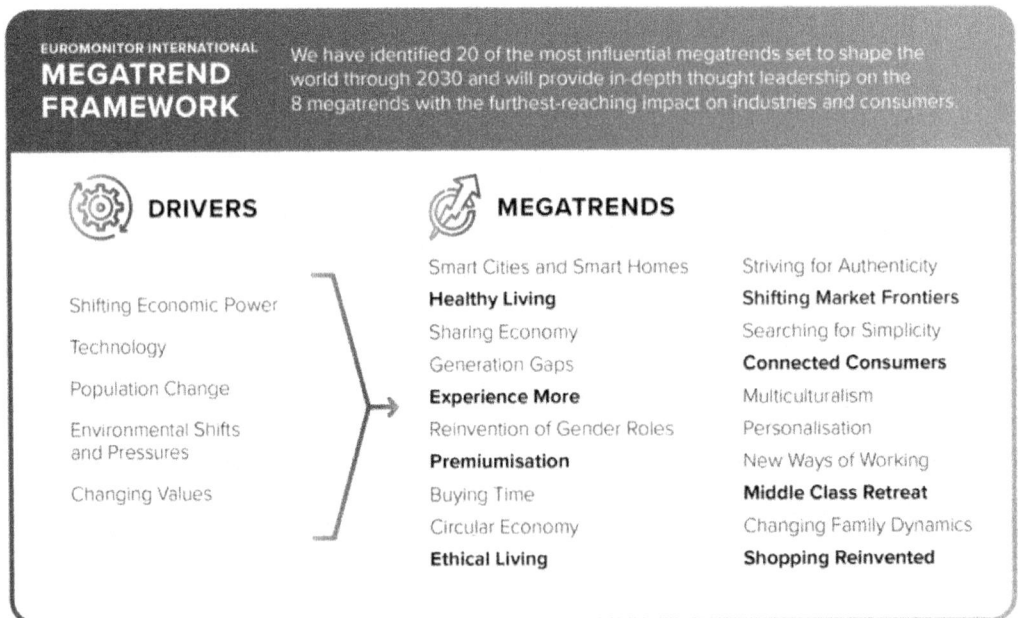

Figure 6.16: Megatrends following COVID-19 (courtesy of Euromonitor International)

Euromonitor International (2020) has identified 20 megatrends that have emerged from the COVID-19 crisis (see Figure 6.16), many of which will influence future travel behavior. These include the need for connectedness, ethical living and healthy lifestyles and the desire to prioritize experiences. Many in the industry believe the pandemic could engender a positive change in traveler behavior. "I think customers will be more aware of the impact of travel on the environment and the communities they visit, and make more considered choices," said Intrepid Travel CEO James Thornton (Dunford, 2020). Thornton believes we will see a renewed focus on slower travel, including train journeys and cycling, as well as keeping experiences as local as possible. Offering more off-season departures are part of Intrepid's post-COVID plans, with wilderness and wellness trips tipped to be of most interest.

The future will be technology-driven

As documented throughout this book, during the COVID-19 lockdown period there was a transformation in the way people interacted with each other, received medical care, spent leisure time, and conducted many of the routine transactions of life. These changes accelerated the migration to digital technologies at stunning scale and speed, across every sector (Sneader & Sternfels, 2020). Robotic technology and artificial intelligence (AI) was already rapidly gaining popularity within the tourism and hospitality industry before COVID-19 (Ivanov et al., 2019), as witnessed by the popularity of Royal Caribbean's Bionic Bars, referred to earlier. But historian and author, Yuval Noah Harari believes that the pandemic can only accelerate the world of automation. "We now see an increase in automatization, with robots and computers replacing people in more and more jobs in this crisis" (dw.com, 2020).

Automation was used in a number of ways by the travel industry during the crisis. Chapter 2 discussed how virus-killing robots were employed to clean hotels in the US, and AI-technology assisted the travel sector with processing refunds for cancelations. Software maker Automation Anywhere Inc., for example, created a bot to help airlines tackle the surge in demand for refunds. "The first airline to implement the bot was able to process 4,000 requests daily, instead of 500, without requiring any additional employee help," said Prince Kohli, the company's chief technology officer (Loten, 2020). On the beaches of Athens, drones were used to enforce social distancing rules as Greece opened up to tourism. Zisimos Zizos worked for the municipality as a photographer, but when the coronavirus hit he swapped his camera for the drone. His job was to prevent large gatherings on the beach to reduce the spread of COVID-19. If beachgoers ignored his call to disperse, the municipal police would step in and could impose fines of around $1,000 per person. "On the busy days, I'll be out here checking on things every few hours," said Zizos (Bouras, 2020).

Finally, the future is likely to see an increase in surveillance technology and digital tools to track travelers – an example being the Known Traveler Digital Identity initiative (KTDI) referred to earlier. Such tools may become commonplace, although Yuval Harari warns that if this surveillance moves from over-the-skin to under-the-skin, it may have dangerous consequences. "Over-the-skin-surveillance is monitoring what you do in the outside world, where you go, whom you meet, what you watch on TV or which websites you visit online. It doesn't go into your body. Under-the-skin-surveillance is monitoring what's happening inside your body. It starts with things like your temperature, but then it can go to your blood pressure, to your heart rate, to your brain activity. And once you do that, you can know far, far more about people than ever before. This could easily lead to the creation of dystopian totalitarian regimes" (dw.com, 2020).

On a final note, despite these warnings, Harari believes that historians in a thousand years will look back at the COVID-19 crisis as a mere bump in the road for the human race. For the travel industry though, the events of 2020 might seem more like a road block. Hopefully this book has provided a few ideas for navigating around this road block, and for being prepared when the next one comes along.

Case study: Surviving the 'new normal'

Figure 6.17: The Sensory and Wit Bar in Canmore (courtesy of Terry Jensen)

Opening The Sensory Restaurant and Wit Bar in Canmore, Alberta in 2019 was a lifelong dream for owner Terry Jensen, who previously ran two McDonald's franchises in the area. Prior to the COVID-19 outbreak, Jensen's two-level upscale restaurant in the Shops of Canmore, with sweeping views of the surrounding Rocky Mountains had been gathering momentum. But in early June, at the start of the 2020 summer season, Jensen was in survival mode as he discussed adapting to the 'new normal'.

Jensen is one of 22,000 tourism-related small businesses in Alberta that between them make up a very large percentage of the tourism sector – a sector that generated over $8 billion for the province in 2019. But the COVID-19 crisis knocked the wind out of most of those businesses, including Jensen's. "It's been stressful for sure but I've been used to change from my old job – with McDonald's. That really teaches you how to adapt and change." At the start of the lockdown Jensen thought about doing takeout but was reluctant as he had not done it before and would be starting with no market share whatsoever. "But then it got to the point that both my wife and I were getting emails saying 'we hear you're closing down'. So we started doing takeout from Monday to Saturday 5-8pm to show people that we were still here." Jensen didn't make any money from this sideline, but felt that it was important to be visible.

When Canadian restaurants were allowed to open up again after lockdown, Jensen said it was just like starting from scratch. "It's tough being our first year, opening up and everyone gradually getting to know your name. At Christmas we started to see momentum and we thought if only we can get to Spring Break, we'll feel good. Now it feels like I'm back to opening up again, but the good thing is that I'm not going to make any of the same mistakes as I did back in April 2019." Jensen had to make changes to meet new regulations as he opened back up, but he also realized he would need to adapt his business concept. "Like everyone else we had to get rid of tables, get masks, hand sanitizer and abide by government cleaning regulations. But the biggest change was that we had two concepts: a fine dining restaurant, The Sensory, upstairs and a bar, The Wit, downstairs with cocktails, fun food, which was just taking off with trivia nights, and live music on Friday nights. Now we don't know if and when the bar element will come back. Also, upstairs our fine dining included three, five and seven course tasting menus which catered to every dietary restriction. But with the new seating limitations we have had to disappoint our chef and go back to just a three course and a la carte menu. It is now 'burn and turn' or we will not survive." In order to adapt to the new 'low-touch' environment, Jensen was planning to get rid of paper menus. "We are going digital over the next two weeks – customers, for example, will access menus via QR codes".

Jensen opened up for evening service only to begin with, offering a hybrid menu served both up and downstairs to make it easier for the kitchen and to be more economical. "We now need a host or hostess every night because of the new regulations – so I have to budget that in. This is one of the new costs of running a restaurant in the new world. I have heard other restaurants doing a 'COVID charge' – a surcharge on the bill. I'm not doing that." Jensen does admit, though, that he has had to change his pricing. "I hope everyone does this. Restaurants in the area were in such a price war that some of them were going to fail. I was doing a $12 lunch but I can't afford that now. I can no longer get into price wars with everyone or I'll go out of business." Jensen was also looking at ways he could cut overheads. "I have found a cheaper reservations company and, before, I paid for a scheduling program but now I do this myself and it costs me nothing."

Jensen has been fortunate in being able to retain his staff. "Except for my manager, 100% of my staff have stayed on which is great as elsewhere I know it has not always been the same. Even if they are getting government support, they are allowed to make $1,000 per month so some are getting both." As for recruiting new employees, Jensen has had mixed success. "I put an ad for a cook and got more applications than I've ever seen for a cook – between 70 and 80 enquiries. At the same time, I put an ad in for a host and haven't got a single application. It goes back to 'why should I host for $15 per hour when I can get government support.'"

While social distancing regulations are still in place, Jensen planned to take advantage of the private room he has. "We have a small event space upstairs – it was for 40 people and that is now 25. So I am advertising that in the local newspaper, and via social media. It has its own private entrance and its own servers who only take care of that party. We are promoting weddings, graduations, parties, rehearsal parties, anniversaries, birthdays. With only 25 guests, a party can be socially distanced with people feeling comfortable but still able to enjoy it. We've been open again for 10 days and in that time I've already booked 10 weddings. This is good, because, for us to survive we are going to have to have a great summer."

Jensen did receive some financial support from the Canadian government during the crisis, which really helped. "We had a $40,000 loan from the government which, if we pay it back by the end of 2021, we get to keep $10,000 of it. Also a waste subsidy of 75% off – we qualified for June and I'm hoping we will for July and August too. There's also a program for landlords from which we've benefited. The landlord gives up 25% of rent, then I pay 25% and the government pays the remaining 50%. I've done this for April, May and June so far. My landlord loves this because, by giving up just 25% for three months, he is assured that his tenants are going to stay. Other landlords around Canmore are not so agreeable which may lead to their tenants leaving."

Despite the government support, Jensen was concerned about the economy, and for him the next six months would be a lot about survival. However, like all successful entre-preneurs, Jensen was moving forward with optimism. "We will continue to offer high end service and quality food to our guests moving forward to set us apart. We want to be 'the place' to eat in Canmore this summer."

Sources: Personal communication with Terry Jensen, June 2020

References

Accor (2020). *ALL Stay Well*. https://www.allstaywell.com/

Adgate, B. (2019). When a recession comes, don't stop advertising. *Forbes*, 5 September. https://www.forbes.com/sites/bradadgate/2019/09/05/when-a-recession-comes-dont-stop-advertising/#3ce95afc4608

Barry, S. (2020). The future of dining out. *Gensler*, 14 May. https://www.gensler.com/research-insight/blog/the-future-of-dining-out

BBC News (2020). Italy open for tourists again. *BBC News*, 2 June. https://www.bbc.com/news/live/world-52900960/page/5

Board of Innovation (2020). *The Winners of the Low Touch Economy*. https://www.boardofinnovation.com/low-touch-economy/

Bort, R. (2020). Can the restaurant industry be saved? *Rolling Stone*, 11 May. https://www.rollingstone.com/culture/culture-features/can-the-restaurant-industry-be-saved-995037/

Bouras, S. (2020). Greece banks on its low coronavirus rate and a long list of new safety rules to lure visitors back to its tourism hotspots. *BBC Worklife*, 9 June. https://www.bbc.com/worklife/article/20200609-how-greece-plans-to-welcome-back-tourists

Broughton, K. (2020). Hotels fund more cleaning by cutting room amenities and breakfast buffets. *CFO Journal*, 26 May. https://www.wsj.com/articles/hotels-fund-more-cleaning-by-cutting-room-amenities-and-breakfast-buffets-11590183440

Buchholz, K. (2020). What could a travel bubble look like? *Statista*, 5 May. https://www.statista.com/chart/21608/travel-bubble-australia-new-zealand/

Byers, J. (2020). Club Med announces "Safe Together" hygiene and safety plan. *TravelPulse Canada*, 27 May. https://ca.travelpulse.com/news/hotels-and-resorts/club-med-announces-safe-together-hygiene-and-safety-plan.html

CBC News (2020). Vietnam, with no recorded coronavirus deaths, craves tourist dollars again. *CBC News*, 19 May. https://www.cbc.ca/news/world/vietnam-coronavirus-tourism-1.5574957

Crabtree, J. (2020). Welcome to a world of bubbles. *Financial Post*, 1 June. foreignpolicy.com/2020/06/01/travel-bubbles-borders-flights-coronavirus-uk-france-air-bridge/

Davies, P. (2020). Many over-50s indicate reluctance to cruise amid coronavirus concerns. *Travel Weekly*, 22 April. https://www.travelweekly.co.uk/articles/368571/many-over-50s-indicate-reluctance-to-cruise-amid-coronavirus-concerns

Del Bello, L. (2020). How Covid-19 could redesign our world. *BBC Future*, 27 May. www.bbc.com/future/article/20200527-coronavirus-how-covid-19-could-redesign-our-world

Donn, N. (2020). Restaurants "without clients or money" to return to 100% capacity. *Portugal Resident*, 31 May. https://www.portugalresident.com/restaurants-without-clients-or-money-to-return-to-100-capacity/

Dowd, V. (2020). How Europe's art world is welcoming back visitors. *BBC News*, 1 June. https://news.yahoo.com/europes-art-world-welcoming-back-003151015.html

6

Dube, K. Nhamo, G. & Chikodzi, D. (2020). COVID-19 cripples global restaurant and hospitality industry. *Current Issues in Tourism*. DOI: 10.1080/13683500.2020.1773416

Dunford, J. (2020). 'Things have to change': Tourism businesses look to a greener future. *The Guardian*, 28 May. https://www.theguardian.com/travel/2020/may/28/things-had-to-change-tourism-businesses-look-to-a-greener-future

dw.com (2020). Yuval Noah Harari on COVID-19: The biggest danger is not the virus itself. dw.com. https://www.dw.com/en/virus-itself-is-not-the-biggest-danger-says-yuval-noah-harari/a-53195552

Euromonitor International (2020). *Coronavirus: Implications on Megatrends.* https://go.euromonitor.com/webinar-consumers-2020-covid-19-impact-on-megatrends.html

Farley, D. (2020). The travel 'ache' you can't translate. *BBC Travel*, 24 March. http://www.bbc.com/travel/story/20200323-the-travel-ache-you-cant-translate

Fox-Lerner, A. (2020). In China, a glimpse at the future of restaurants. *Eater*, 24 March. www.eater.com/2020/3/24/21191278/china-beijing-coronavirus-future-of-restaurants

Friedman, R. (2020). Open for business: Hotels post-COVID-19. *Commercial Property Executive*, 4 May. https://www.cpexecutive.com/post/open-for-business-hotels-post-covid-19/

Future Cruise (2020). *An Industry in Lockdown.* May. https://future-cruise.nridigital.com/future_cruise_may20/issue_12

Genung, A. (2020). Here's what eating out might look like when restaurants reopen. *Eater*, 21 April. https://www.eater.com/2020/4/21/21229650/what-eating-out-might-look-like-when-restaurants-reopen-coronavirus-impact

Girneata, A. (2014). Adaptability – a strategic capability during crisis. In J.T. Karlovitz, *Economics Questions, Issues and Problems*, Bucharest, Romania, pp. 243-149.

Global Data (2020a). Cruise industry faces uphill battle to attract customers post-COVID-19. *GlobalData*, 3 May. https://www.globaldata.com/cruise-industry-faces-uphill-battle-to-attract-customers-post-covid-19-says-globaldata/

Global Data (2020b). *Coronavirus (COVID-19) Executive Briefing.* Global Data PLC. https://globaldata.com/covid-19/

Glusac, E. (2020). Hotels vs. Airbnb: Has Covid-19 disrupted the disrupter? *The New York Times*, 14 May. https://www.nytimes.com/2020/05/14/travel/hotels-versus-airbnb-pandemic.html

Hamdi, R. (2020). Marketers confront what travel will look like post-crisis. *Skift*, 6 April. skift.com/2020/04/06/marketers-confront-what-travel-will-look-like-post-crisis/

Hancock, C. (2020). Covid-19: Temperature checks for guests planned for Fitzpatrick Hotels in New York. *The Irish Times*, 5 May. https://www.irishtimes.com/business/transport-and-tourism/covid-19-temperature-checks-for-guests-planned-for-fitzpatrick-hotels-in-new-york-1.4244826

He, H. & Harris, L. (2020). The impact of Covid-19 pandemic on corporate social responsibility and marketing philosophy. *Journal of Business Research*, **117**, 176-182.

Heath, R. & Jin, B. (2020). Ranking the global impact of the coronavirus pandemic, country by country. *Politico*, 21 May. https://www.politico.com/interactives/2020/ranking-countries-coronavirus-impact/

Hudson, S. & Hudson, L.J. (2017a). *Customer Service for Hospitality & Tourism*. Second Edition. Goodfellow Publishers Limited, Oxford, U.K.

Hudson, S. & Hudson, L.J. (2017b). *Marketing for Tourism, Hospitality, and Events*. Sage, London.

IQ Latino (2020). COVID-19: This summer in Spain beaches will be divided into sectors, time slots and capacity will be limited. *IQ Latino*, 13 May. https://iqlatino.org/2020/covid-19-this-summer-in-spain-beaches-will-be-divided-into-sectors-time-slots-and-capacity-will-be-limited/

Ivanov, S., Gretzel, U., Berezina, K., Sigala, M. & Webster, C. (2019). Progress on robotics in hospitality and tourism: A review of the literature. *Journal of Hospitality and Tourism Technology*, **10**(4), 489-521.

Jones, S. (2020). Balearic Islands prepare to welcome 11,000 German tourists. *The Guardian*, 14 June. https://www.theguardian.com/world/2020/jun/14/balearic-islands-prepare-to-welcome-11000-german-tourists

Liang-Pholsena, X. (2020). Does a generational divide await Asia Pacific's return to travel? *Skift*, 13 May. https://skift.com/2020/05/13/does-a-generational-divide-await-asia-pacifics-return-to-travel/

Lockhart, K. (2020). Life after lockdown in Vietnam: This is what it's like when an entire country reopens. *CNN Travel*, 15 May. https://www.cnn.com/travel/article/life-after-lockdown-vietnam-domestic-travel/index.html

Loten, A. (2020). Travel industry automates pandemic response with new digital tools. *Wall Street Journal*, 3 May. https://www.wsj.com/articles/travel-industry-automates-pandemic-response-with-new-digital-tools-11588361276

Marchese, D. (2020). David Chang isn't sure the restaurant industry will survive Covid-19. *New York Times*, 27 March. https://www.nytimes.com/interactive/2020/03/27/magazine/david-chang-restaurants-covid19.html

Murphy, C. (2020). The future of conference: Will events remain virtual after lockdowns? *USA Today*, 9 May. https://www.usatoday.com/errors/404

Newstead, A. (2020). Drive-in concerts are a thing now. We spoke to the musician who performed the world's first. *ABC Music News*, 7 May. www.abc.net.au/triplej/news/musicnews/drive-in-concert-coronavirus-danish-mads-langer-interview/12223420

Nguyen T.M. & Malesky, E. (2020). Reopening Vietnam: How the country's improving governance helped it weather the COVID-19 pandemic. *Brookings*, 20 May. https://www.brookings.edu/blog/order-from-chaos/2020/05/20/reopening-vietnam-how-the-countrys-improving-governance-helped-it-weather-the-covid-19-pandemic/

Northstar (2020). Survey shows improved confidence in live events. *Northstar Meeting Group*. https://www.northstarmeetingsgroup.com/News/Industry/Weekly-Survey-Tracking-Coronavirus-Meetings-Response

6

Nunn, J. (2020). Restaurants will never be the same after coronavirus - but that may be a good thing. *The Guardian*, 14 April. www.theguardian.com/commentisfree/2020/apr/14/coronavirus-restaurants-pandemic-workers-communities-prices

Posaner, J., Cokelaere, H. & Hernandez-Morales, A. (2020). Life after COVID: Europeans want to keep their cities car-free. *Politico*, 11 June. https://www.politico.eu/article/life-after-covid-europeans-want-to-keep-their-cities-car-free/

Puckett, J. (2020). How airports will change after COVID-19. *Conde Nast Traveler*, 7 May. https://www.cntraveler.com/story/how-airports-will-change-after-covid-19

QUO (2020). *Podcasts on the Future of Travel*. https://www.quo-global.com/podcasts/

Quy, N. (2020). Vietnam among the first economies likely to restart international tourism after the pandemic. *Vietnam Insider*, 27 May. https://vietnaminsider.vn/vietnam-among-the-first-economies-likely-to-restart-international-tourism-after-the-pandemic/

Reeves, M & Deimler, M. (2011). Adaptability: The new competitive advantage. *Harvard Business Review*, July/August. https://hbr.org/2011/07/adaptability-the-new-competitive-advantage

Ritter, K. (2020). Disinfected dice: Las Vegas casinos getting ready to roll. *CTV News*, 22 May. ttps://www.ctvnews.ca/world/disinfected-dice-las-vegas-casinos-getting-ready-to-roll-1.4950529

Rizzo, C. (2020). Airbnb releases new cleaning protocol program in wake of coronavirus. Travel + Leisure, 3 June. https://www.travelandleisure.com/hotels-resorts/vacation-rentals/airbnb-host-coronavirus-cleaning-training

Romei, V. (2020). Tourism deals lingering blow to global economy. *Financial Times*, 13 June. https://www.ft.com/content/b406133d-5859-4f68-8163-b179f50f6a22

Saladino, 2020). Bars and restaurants have a chance to change for the better. Will they take it? Winemag.*com*, 16 May. https://www.winemag.com/2020/04/28/will-restaurants-reopen/

Santoro, A. (2020). The proposed guidelines Disney World will follow once the park reopens in July. *Yahoo Sports*, 26 May. https://sports.yahoo.com/disney-look-upon-reopening-see-225237253.html

Sapient, P. (2020). Preparing for a post COVID-19 world: 4 ways travel brands can learn from past events. *Skift*, 9 April. https://skift.com/2020/04/09/preparing-post-COVID-19-world-travel-brands-learn-past-events/

Shaffner, L. (2020). Live Nation CEO outlines plan to resume concerts + tours. *Loudwire*, 9 May. https://loudwire.com/live-nation-ceo-michael-rapino-resume-concerts-coronavirus/

Simon, E. (2020). CBRE hotels research: Full demand recovery by late 2022. *Hotel Management*, 21 May. https://www.hotelmanagement.net/operate/cbre-hotels-research-full-demand-recovery-by-late-2022

SimpliFlying (2020). *The Rise of Sanitized Travel. The Day of the Life of an Airline Passenger.* April. https://simpliflying.com/guidance-airlines-covid-19/

Singh, S. (2020). Hotel design may undergo significant changes post Covid-19 pandemic. *ET Hospitality World,* 14 April. https://hospitality. economictimes.indiatimes.com/news/operations/architecture-and-design/ hotel-design-may-undergo-significant-changes-post-covid-19-pandemic/75140021

Singh Maini, T. (2020). Vietnam's success in dealing with the Covid-19 pandemic. *Modern Diplomacy,* 16 May. https://moderndiplomacy.eu/2020/05/16/ vietnams-success-in-dealing-with-the-covid-19-pandemic/

Sneader, K. & Sternfels, B. (2020). From surviving to thriving: Business after coronavirus. *McKinsey & Company,* 1 May. www.mckinsey.com/featured-insights/future-of-work/ from-surviving-to-thriving-reimagining-the-post-covid-19-return

Taub, A. (2020). Why are women-led nations doing better with Covid-19? *New York Times,* 18 May. https://www.nytimes.com/2020/05/15/world/coronavirus-women-leaders.html

Tatarski, M. (2020). Vietnam halted Its COVID-19 outbreak. Now comes the economic fallout. *World Politics Review,* 18 May. www.worldpoliticsreview.com/articles/28770/ in-vietnam-economy-could-be-hit-hard-despite-an-effective-covid-19-response

UNWTO (2020a). *COVID-19 Tourism Recovery Technical Assistance Package.* https:// webunwto.s3.eu-west-1.amazonaws.com/s3fs-public/2020-05/COVID-19-Tourism-Recovery-TA-Package_8%20May-2020.pdf

UNWTO (2020b). UNWTO launches global guidelines to reopen tourism. *UNWTO press release,* 28 May. https://www.unwto.org/news/ unwto-launches-global-guidelines-to-reopen-tourism

Wainwright, O. (2020). Smart lifts, lonely workers, no towers or tourists: Architecture after coronavirus. *The Guardian,* 13 April. www.theguardian.com/artanddesign/2020/ apr/13/smart-lifts-lonely-workers-no-towers-architecture-after-covid-19-coronavirus

6

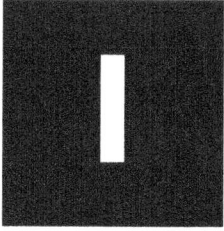

Index

Printed by Printforce, United Kingdom